CORNERSTONES

CORNERSTONES

Wild Forces That Can Change Our World

Benedict Macdonald

BLOOMSBURY WILDLIFE
LONDON · OXFORD · NEW YORK · NEW DELHI · SYDNEY

BLOOMSBURY WILDLIFE
Bloomsbury Publishing Plc
50 Bedford Square, London, WC1B 3DP, UK
29 Earlsfort Terrace, Dublin 2, Ireland

BLOOMSBURY, BLOOMSBURY WILDLIFE and the Diana logo are trademarks
of Bloomsbury Publishing Plc

First published in the United Kingdom 2022

A catalogue record for this book is available from the British Library.

Library of Congress Cataloguing-in-Publication data has been applied for.

ISBN: HB: 978-1-4729-7160-9; ePub: 978-1-4729-7156-2;
ePDF: 978-1-4729-7158-6

2 4 6 8 10 9 7 5 3 1

Typeset in Bembo Std by Deanta Global Publishing Services, Chennai, India
Printed and bound in Great Britain by CPI Group (UK) Ltd, Croydon CR0 4YY

To find out more about our authors and books visit www.bloomsbury.com
and sign up for our newsletters.

To my dear cousin Timothy Chiles, who taught me the value of all living things in nature, and what they do, from a very early age.

Contents

Contents

INTRODUCTION

We stepped through the security fence, and the chilling sameness of our countryside gave way to sudden, riotous variety. The birch woodland was filled with grasses, sedges and bramble, but the bramble had been neatly clipped into thorny islands by some unseen gardener. The birdsong was turned up – sudden and loud. A marsh tit buzzed in a stand of newly coppiced willows whilst blackcaps and garden warblers bubbled and fluted alongside. As we walked through the woodland, some fallen trees already had new saplings, and fungi, pushing up through their decaying forms. Water lay like a film across the forest floor. Soon, we reached the first of the ponds.

Just three years earlier, the landscape we were looking at now had been a straight river – a tributary of the Tinney, near Ladock in Cornwall; an area of Britain prone to increased winter flooding in recent years. Now, in place of a straight torrent, a calmed wetland lay within the woodland. A kingfisher glowed on a young birch, watching the pond with intent, which, upon closer inspection, skipped with young brown trout that repeatedly broke the surface. The pig-like screech of a water rail carried from the sedges. All around the edges of the impounded river were bushlands of the kind rarely seen in our country; dense, billowing willow bushes, like perpetually frozen explosions, mingled with birches. A large oak had been felled close by and lay across the edge of the pond, creating a frenzy of microhabitats in its fallen wake; little wood-lined ponds, bramble tunnels and nettles. The water-loving alders, alone, grew straight; they had largely been left intact. With a sad little flute, a male

bullfinch flew with his mate across the pond, their white
rumps flashing as they vanished into another corner of their
shade. The sunlit edges of the newly formed pond were
silvered with a lithe new crop of common frogs. Each reed
appeared to harbour a dragonfly: broad-bodied chasers and
common hawkers cruised around.

There was a hushed silence amongst the group. One
remarked that it felt as if we had stepped into a Canadian
wilderness. The landowner, a respected local beef farmer,
Chris Jones, appeared delighted. He led on, pushing ever
deeper into this Cornish wild. 'Look', he observed, 'how fast
the new willows have grown back.'

We had stopped by a fallen branch, and yet it was not
quite detached from the tree. Precise carpentry had chiselled
away at the trunk to leave a triangular, spear-like point.
Here, the rest of the willow had fallen into the water, but
already new shoots were sprouting, like feathered tentacles,
from the place where the cut had been made. All around the
pond's edge, previously straight-growing trees had been
reformed into bushes, with new branches bursting out of
every hewn limb. Deadwood and live shoots now existed
side by side. At one point, close to the bank, we heard a
deep, low growl from within the earth, as if the soil itself was
groaning. We hurriedly stepped back – and left the unseen
inhabitant in peace, below.

As we ventured on, I mentioned to Chris that this was
surely the ideal habitat for the UK's most rapidly declining
resident bird – the willow tit. An inhabitant of dense, damp
scrub-woodland, which excavates its nest hole in only the
softest of standing timbers, the willow tit has vanished from
many of its former haunts. Within my lifetime alone, I have
watched entire populations disappear from areas close to
home. But then, with a dramatic sense of timing, I heard an
angry, nasal, 'tzchay, tzchay, tzchay,' – and there were a pair of
willow tits, moving restlessly through the new willow scrub
as if no one had told them they were rare. It would emerge

later that a small population still survived nearby, and from here they had been able to colonise the newfound chaos. A little while later, we found an uneven, small hole – like that of a tiny, untidy woodpecker – dug into one of the decaying willow stumps by the water's edge: the willow tit's distinctive nest hole. And their habitat, like that of the kingfisher and water rail, the brown trout and common frog, had been magicked into being in just three years.

Evening was drawing in, and the first noctule bats appeared to hawk low over the pond. The air above the pond's surface danced with small, airborne insects. As the colours deepened and the willows turned to shadows, we caught the distinctive paler form of a Daubenton's bat as it whisked low over the water. Now, the scene was set. We had not come here for kingfishers or willow tits, bats, frogs, fish or fungi. We had come to see those who had created and reformed this landscape.

The water was a vivid deep orange, in the late evening night, when the first beaver appeared. A female, she moved with great grace, leaving the smoothest of wakes behind her. At one point, someone on the bank moved a little too suddenly for her liking; with a tail-slap she was gone, but soon resurfaced again. Settling into her routine, the female beaver proceeded to the felled oak, where, from the safety of the water, she began to beaver away; zealously hewing away at a large branch, snapping it off, then holding it deftly between her claws. At this point, three far-smaller wakes appeared from the opposite side of the pond, and her three kits came to join her. Hearts were collectively melted as it was realised that each kit was carrying a tiny beaver-twig of its own: beavers learn their craft from a very early age.

The three kits joined their mother in the water below the fallen oak. This sight would, prior to the 1300s, have been common across southern England. There was enormous poignance, therefore, in knowing that we were the first generation in almost 700 years to witness it again – and to

stand within a southern English landscape whose very
character and wildlife had been shaped, above all, by the
actions of one animal alone. More so that this mammal was
not, for once, ourselves.

Over the course of the next hour, the female beaver and
her kits, which had perhaps left the lodge just a week before,
continued to hew and nibble away at the branches of the
fallen oak, and eventually the male came to join them. For
15 minutes, the male and female groomed one another,
delicately removing stubborn twigs from their lustrous,
felted fur. The kits eventually became more adventurous,
setting out across the pond in search of new, tiny twiglets of
their own. The air was now thick with moths as bats sallied
into their midst above the beaver-made pond. As the sun set,
we left the beavers to do what they do best.

The following morning, it became clearer to what extent
the entire landscape here – enclosed for now, within a fence –
had been reformed by beavers. Not one, but a series of dams
had transformed a narrow tributary – of relatively limited
use to many native species – into a complex, tiered set of
impounded ponds, loud with birds and rich in fish. The
ponds, Chris explained, were acting as important crèches for
young trout, which, with surprising ease, then used 'fish
passes' around the dam; small areas where the river successfully
breaks free of each beaver dam to continue its course, calmed
and slowed, downstream. Over the past three years, the vast
majority of willows and other trees gnawed by beavers – up
to 80 per cent – had survived, evolving into ever denser
coppiced forms. Now, these were home to a range of
birds that cannot make use of live, straight trees – from reed
buntings to nesting garden warblers. The vegetation
complexity of the forest floor was striking; whilst beavers are
well known for felling branches, they also coppice bramble,
creating nesting 'islands' for birds, and refuges for nectaring
insects, whilst not allowing this fast-spreading bush to
rampage throughout the woodland. The dams themselves

were like trellises: ox-eye daisies, meadowsweet, rosebay willowherb and purple loosestrife were bursting up through them. A jay had perhaps passed by and buried some acorns, as a few small oaks were also beginning their journey upwards from the fertile, silted base of the dam, into the light.

The Cornish beaver project's enclosure is a mere two hectares in size: a microcosm of what beavers can do to a landscape, of the life they can bring, and the ways in which they can protect and restore our waterways. As we left, Chris, a lifelong farmer, reflected on what these ponds meant to him. 'As a farmer, we know there are floods ahead, and we know there will be summers of drought. If we are to survive, to water our crops, to keep our livestock on the land, we must keep slow water on the land. It's the most essential thing of all – and so, in truth, is the beaver. Beavers are not only solutions to this problem – they may be the *only* solution. It's not just about wildlife, or natural wonder. We need these species to survive.'

You might think of an ecosystem as a habitat – populated by the animals that live *within* it. Beavers, after all, live beside rivers. Lions roam the plains of Africa. Whales swim in the ocean. Trees grow within forests. In recent decades, however, ecologists have begun to realise that this is not entirely how things work. Instead, certain flora and fauna actively shape and create the very ecosystems in which they and others live. These have become known as ecosystem engineers – or keystone species. In this book, I refer to such powerful wild species as 'cornerstones': invaluable but often missing components of our depleted natural world – something especially true here, at home in Britain.

While every living organism has a niche, an influence on the world around it, and – it might be argued – a right to survive, not every animal is equal in the *effect* it brings to bear upon the world. A puffin, for example, can feast within a zooplankton-rich fishery, created and maintained by great whales – but whales do not require puffins to survive. An otter can fish within a beaver's complex network of ponds – but otters do not prescribe the world in which a beaver lives. A beaver, by contrast, re-forms entire landscapes, new worlds in which other animals, including otters, find food or a home.

In recent times, the once-powerful effect of cornerstone species has been dramatically reduced. Now, the most powerful ecosystem architect walks on two legs, re-forming the world in its image. Yet we humans are far from the only architects of our planet's ecosystems. There are others who just as surely shape and enhance the natural world, if we allow them to.

Some of these species are multitudinous or familiar, like the bees or trees in our gardens. Others are so distant to our memories as to be folklore: the wolves that once shaped our woodlands or the great whale pods, whose frothing plumes were familiar sights to those living on our island three centuries before us. Yet however long these species have been lost, the laws of ecosystems have not changed, and even now, the impact of their removal reverberates through our impoverished landscapes and seas.

In Britain, we watch catastrophic flooding in our uplands, forgetting the beavers that once impounded and slowed those very upland streams. We erect expensive fences and plastic tubing to protect trees against deer in the Highlands of Scotland, forgetting that wolves are the surest tree guardians of all. We have forgotten the power of wilder forces to enhance the world around us – a world of which we, too, are an intrinsic living part.

It is now widely accepted by conservationists that the loss of cornerstone species over centuries, if not millennia, has profoundly impoverished the world's ecosystems, as it has ours here in Britain. Even now, our contemporary flora and fauna reveal adaptations to giant herbivores now absent from our shores. Yet in terms of our ecology, the departure of such animals is but a recent event. Fortunately, many of the world's ecosystem architects have survived. And we can now restore them – should we wish.

Some of these species, like the wolf or the lynx, are now recovering in many parts of Europe following centuries of persecution, but are yet to return to Britain. Others, like the beaver or boar, are beginning a long journey towards acceptance here on our own shores. Some giants, like the great whales, are venturing ever more frequently into our waters. Eagles are returning to British skies not darkened by their presence in centuries.

Meanwhile, it is perhaps ironic that bees, small creatures of the soil, and apparently innocuous yet invaluable trees, such as rowan or cherry, are vanishing quietly all around us. All the while, that most dominant keystone species of all – ourselves – grows ever more disproportionate in its effect upon the British landscape.

At a global level, many have called for the restoration of living systems as the most powerful of tools to protect our climate – and save our species from the Sixth Mass Extinction. In the words of Sir David Attenborough, 'We must rewild the world.' And to do just that, we must restore its stewards.

In recent years, a growing movement to restore lost landscapes and return lost species has emerged here in Britain, too. Yet despite a few pioneering projects, our nation still struggles with the concept of ceding power, or responsibility, to any animal except ourselves. Indeed, many British people have perhaps become so accustomed to a landscape shaped by us, and manicured *for* us, that it can be hard to accept the critical role played by others who once shaped our native landscapes. For some of us, wild animals are not ecosystem imperatives but optional additions to our landscape: luxuries at best and nuisances at worst. It is not yet understood that returning such animals will also create a richer world for *us*. But slowly, things are beginning to change.

We are beginning to see beavers transform our waterways, the powerful effect of free-roaming large herbivores on rewilded lands, and the transformative grace of woodland regeneration – all here, on our doorstep, in Britain. More and more groups are calling for the reintroduction of the lynx. And slowly, in spite of our species' extreme reluctance to cede power, our minds are slowly expanding – and so are the possibilities of nature restoration in our country.

In the pages that follow, *Cornerstones* takes you deep into the lost world of Britain's native keystone species. It invites you to wander with wild architects; to share in the riot of life they create, and to walk, swim or fly within the worlds they engender. Living beside them, you may discover a growing admiration for their ways, a more profound sense of wonder at the richness they create. In time, it may become easier for you, as the dominant species on our island home, to cherish and live beside our fellow stewards once again.

CHAPTER ONE
Boar

In my early school days, July heralded the start of the holidays as my parents released me into the flower-filled glades of the Dordogne. Here, I could be safely relied upon to run, not in a straight line, but in excited circles, chasing a range of fairy-tale butterflies long lost in numbers to Britain. The bramble-filled edges of the shaded forests flickered with wood whites and white admirals. Heath fritillaries became commonplace distractions as they alighted on scabious. The ethereal black-veined white, a spectral ghost of the British countryside, was common – and sometimes, beating me for pace, a large tortoiseshell would lollop through, tauntingly alighting on a high oak leaf before bouncing away into the distance. Once or twice in a long day's chasing, all hell would break loose as a purple emperor came sailing through, landing on a dung-pile at a clearing edge, or a majestic swallowtail sailed across the meadow, rarely deigning to stop for admiration.

The Dordogne's woodlands were broken with meadows, each home to an abundance of butterflies I would never see back home. But the richest hunting grounds were often the most unlikely. Often, my long-suffering parents would bring the car to a halt beside a patch of wasteland, a roadside verge or an unpromising area of rubble as I glimpsed from the window the buttery shades of a Berger's clouded yellow or the cobalt of a baton blue. These little rocky wastes, with loose soils and bursting clusters of wildflowers, often held areas of extremely high butterfly diversity. Over time, over successive holidays, this little mystery became more and more intriguing. Why did these disturbed areas often hold such a variety of species?

On opening the Collins *Butterfly Guide*, months later, magnifying glass poised over printed film exposure, I would ponder the illustrations by Richard Lewington to separate my fritillaries and brush up on my blues. Having worked out what I had photographed, my curiosity about that particular butterfly would draw me into the dense, terse text at the back of the book. Again and again, two words jumped out from the habitat descriptions: 'disturbed ground'.

The following summer, the quest to discover new species became so extreme that I would be sometimes seen running into the earthy, rubbly edges of supermarket car parks in southern France, where, again and again, new butterfly species would unfailingly be found. Then, one day, we were walking through the more serene surroundings of the Dordogne forest meadows when we chanced upon an earthwork. Like the waste soils of the roadsides, it was bristling with butterflies. Birds-foot trefoil, gentian and cow parsley were all pushing up through the disrupted soil, each flower head flexing with fritillaries. The glistening soil was carpeted with wood whites, their long proboscises extracting nutrients from the rotavated earth. It seemed that we had come across 'disturbed ground'. Once more, here were the butterflies. But who had made the earthwork?

Shortly after, with a sombre, alarming lack of warning, a hunter, or *chasseur*, appeared silently beside us, clutching a large and powerful rifle. He looked disconsolate, having had an unproductive day, and stopped for a chat. My father explained in French that we were chasing butterflies. The hunter looked bemused. He had sharp ears and sharp eyes. '*Ecoutez, la tourtourelle*' (Listen, the turtle dove), he said, as we heard a faint purring from deep within the woods. Then, we pointed out the earthwork. His expression looked disappointed. '*Oui, c'est le sanglier. Mais, c'est vieux.*' And with that, he departed into the forest as silently as he had come.

My first encounter with the Dordogne's *sangliers,* or wild boar, came some days later – in the form of a casserole. Yet whilst their earthworks were to be seen everywhere, in various forms – fresh from the night before or long overgrown with flowers – the animals themselves were as ghosts. It grew to fascinate me that the giant hogs once hunted by Henry VIII and capable, I was told, of 'killing knights', could be watching us from all around. I was unsure about their ambush technique and took great care when approaching the woodland edge. Yet it would be 15 years before I finally set eyes on these mythical beasts – and not in the Dordogne, but far closer to home, in the Forest of Dean.

It was 1 July 2010. The woodland was dense, and the line of sight was broken. There was a ditch ahead, and there was something large within it. I couldn't see until I was very close indeed that the creature was wiry and hairy. The ditch was also far deeper, and thus the animal far larger than I had thought. Then, the creature emerged and raised its head.

There is an instant, primal *punch* in coming face to face with a wild boar. The ancient, wizened, kind, curious face. The massiveness of the towering head. The mossy aspect, as if hewn from a fallen stump. Boar have a presence that projects beyond the physical. They are as ancient as the trees, and their appearance in a woodland I had known for years felt like a temporal passage to another realm. As I watched, the mother, who could smell me, but see me rather poorly, revealed, in her snuffling wake, a sounder of stripy humbug piglets.

At this point, I faced a difficult decision. I was in the course of making my first film, and so, instructing the family group to 'wait there', I dashed back through the woodland to my car. Shouldering my camera and brutally heavy tripod,

I began to make my way back to the mother and her young. The soil was a chaotic mess as if a gardener had just put down his spade, but the boar were nowhere to be seen. The adrenaline compensating for the growing groove in my right shoulder, I, too, ploughed on, following the fresh diggings, one of nature's most obvious mammalian tracks.

It took me almost an hour and a kilometre's worth of walking before I caught up, once more, with the mother and her bouncy, wriggly, rubbery piglets. Then my heart skipped another beat. Staring at me from the shadow of an oak was a *giant*. A male boar weighs up to 100kg – and this one looked every gram of that. His two tusks, and the silvery hairs around them, gave him an austere expression, yet the placid eyes were calm. The boar stared at me with mild curiosity. I stepped back a few paces and slowly sat down. He watched but didn't come any closer. By reducing my profile, it seemed that I appeased him. Males play only a passing role in parenthood but can still defend sows and their young. But soon, his stomach got the better of him, and the boar joined the mother and her humbugs rootling away in the soil.

I sat riveted for an hour as the family industriously turned over more soil than a human gardener might aspire to in a week. The Dean's beloved bluebells bulbs were unceremoniously uprooted – and snorted down. Boar love bulbs and tubers, yet whilst their short-term effect on these lovely plants is to remove them, their longer-term effect is the opposite. Areas rootled by boar enjoy higher rates of bluebell germination, and in all their time in the Dean boar have, perhaps unsurprisingly, given their co-evolution beside bluebells, never wiped the bluebell out.

After soil-snuffling for quite some time, the enormous male sloped away into the forest. As he did, one of the humbug piglets made a run towards me, and the mother followed. I was by now rooted deep behind – and almost within – my tripod, tucked against the base of a tree. As I adjusted the lens, I realised the piglet was now too close for the camera to focus on. Soon, two, three, four, then all seven of the piglets discovered that the most appetising and fertile area to be rootled was now directly below my tripod. Before I could move, I was inundated with snuffling piglets. Then the sow came back to get them – and paused, just a couple of metres away. I could see every toothbrush bristle on her silvered head.

Like any good mother, she was happy as long as her children were in earshot, eyeshot, behaving and content. My being very still and very low, she even now didn't see me as a threat. And once the piglets had been happily established within inches of my shoelaces, she, too, began rootling away, her giant snout a spade to their tiny trowels. Often, the humbugs would move in the mother's wake. She disrupted: they ate. The largest of the litter had a large mound of clay soil stuck firmly to the top of his snout. No matter how much he tried to get it off, it wasn't going anywhere. For the next two hours, the boar shared with me an insight into their industry, gentleness and family kinship that I will never forget. After exhausting the area around the tripod, they finally moved away – and at last, I was able to focus the camera and commit these remarkable creatures to film. Then, suddenly, a branch snapped in the nearby forest. Startled, the female grunted, and she and her humbugs pelted off into the woods. Within five seconds, I had lost them altogether – and this is how most people experience boar: most of the time, you hear a grunt and watch a curly tail vanish into the bracken – leaving only an earthwork in its wake.

Three years later, I was passing the same spot when a flash of purple caught my eye. Glowing in the dappled shade at the woodland edge, it was a vivid cluster of common dog violets.

As I drew nearer, some orange petals grew airborne and fluttered around: a colony of small pearl-bordered fritillaries. They glided and bounced through the sunlight, alighting to nectar on the flowers – the same on which their caterpillars were raised. My mind was transported back three years to the snuffling piglets, and back again to the flower-filled earthworks of the Dordogne. It was another curious case of the boar and the butterflies.

There is nothing like a boar or close to a boar (or indeed, a pig). So successful and versatile has the design proven over time that speciation amongst wild pigs has been incredibly low. Indeed, with the exception of the bearded pigs native to islands such as Borneo, Java, Sumatra and the Philippines, the entirety of Eurasia is ruled by just one species – the wild boar, *Sus scrofa*. Compared to other European animal families such as cats (which have diverged into forms as diverse as the lynx, wildcat and leopard) or canids (in Eurasia, the grey wolf, golden jackal, dhole, raccoon dog and several species of foxes), the boar stands out as an ecological *singularity*.

Boar can be found in the middle of marshlands in Poland and Belarus and in the reed-beds of the marshes in Iraq. They are competent swimmers whose capabilities exceed most of our own: in 2013, one animal made it from mainland France to the island of Alderney, a journey of 11km. Observers in the Black Sea have videoed piglet shoals following in the wake of nautical sows, proceeding to cause mild but comedic havoc amongst beach-goers as they snuffle past deckchairs on their way inland. There is something strangely elegant about swimming boar; they glide through the water with surprising speed and barely a splash.

Boar are equally at home in baking, arid environments, such as the dry *maquis* of the Mediterranean islands or the ancient grassland steppes of Hungary's Hortobágy. Yet few animals on Earth are also more resistant to the cold. In Amurland in Siberia, long after the red deer have migrated to the south, the powerful Amur tiger is left with only prey hardy enough to resist temperatures that can drop below -40°C. Boar may be slowed down by the snow, but they are not deterred by it. Far hardier than brown bears, they do not hibernate, and their powerful snouts can pierce deep snow-cover to dig up food in the depths of winter. Boar, therefore, would originally have formed and reshaped the land in almost all of Europe's habitats, and not just its woodlands. From freezing slopes to warm marsh water, boar can invariably find a way to endure – and force passage.

Evolved in an ecosystem originally dominated by giant grazers and their predators, the boar's role is different and unique: it is the bulldozing digger of the animal kingdom. Whilst wallowing bison can cause significant disruption to the soil, and ponds can form in their wake, no other animal turns over so much soil, with so much consistency – and finesse – as the boar. Each boar can rootle, on average, 40m^2 in the course of a week.

Boar are uniquely attuned to the soil – and to what lives within it. Grubbing mostly for rhizomes (subterranean plant stems), tubers, roots and bulbs, they overturn the surface layer of the soil – but unlike crude human diggers, they do not rupture deep into the earth, and thereby avoid damaging the vital fungal circuitry by which plants and trees communicate. As with beavers, what boar *do* is simple: they turn over soil to uncover their food. The *results* of what they do are myriad and complex.

By rotovating only the top layer of the soil but then moving on to fresh digging grounds, boar act as nature's oldest rotational farmers. As they move, they 'reset' tracts of land back to complex mounds of earth. Unlike a plough, which wipes an entire field back to earth, the diggings of boar are nuanced, textured and three-dimensional. Indeed, any of Britain's 27 million gardeners might recognise this fresh, earthen 'mess' as the basis of their own beloved flowerbeds.

Working within woodlands and clearings, boar rotovate and smash bracken-filled forest floors back into nuanced, ridged environments. Because boar have returned recently to some large British woodlands, it has now proven possible to observe the effects. In 2020, I returned to check a sparrowhawk eyrie in the Forest of Dean, one I had visited 10 years before, finding my way back to it using photos I had taken in the June of that year. The vista at that time had been uniform and dreary: an endless carpet of fern. Bracken is not devoid of life but is profoundly limiting for many species if growing in dense, unbroken carpets. Larger invertebrates such as butterflies or bees cannot stop off in these areas to feed, and a profusion of bracken can often indicate a profusion of the animal that has eaten all else: deer. But, 10 years later, this arid forest floor had been transformed into a woodland garden.

Foxgloves, which, like most wildflowers, flourish in disturbed soils, were bursting through the boar-rooted ground; large red-tailed bumblebees droned through the woodlands, and honeybees were raiding every flower. On closer inspection, a host of other annual plants – sweet woodruff and wood sage, anemones and primroses – were to be found on the upward-facing ridges of the boar diggings; exposed to the glances of the sun.

Over the years in the Forest of Dean, as boar have created more and more nuances in the soil and continually 'reset' areas of forest floor with their spade-like noses, floral diversity has continued to increase. In those areas rooted over several years, common spotted and green-winged orchids have emerged. Orchids take time to disperse and propagate within the soil, but once they do, species like the autumn lady's tresses – whose spiralling helix of white blooms flowers by the very end of summer – can grow to profusions as thick as grass in the boar-tilled gardens of the Dean.

As any human gardener will notice, as soon as fresh soil has been turned over, a range of birds, specially evolved to exploit fresh soil, appear as if by magic. Most of us with gardens can still enjoy the appearance of a robin as soon as we've turned over fresh soil. Swooping onto fresh earth, robins can be conjured as if from thin air as they feast on newly flushed insects and worms. In Britain, most robins now follow spade-, hoe- and fork-wielding primates around. Yet in most of Europe, the shyer, woodland-dwelling robin follows a far more disruptive gardener. Sitting quietly in the Forest of Dean, too, I have often watched robins, generally followed by dunnocks, blackbirds and song thrushes, foraging in the wake of families of boar, the animal they followed long before we tilled our gardens – or indeed colonised the British Isles.

Studies carried out in the French Sologne show that the effects of rootling on woodland birds are not limited to feeding alone; more ground-nesting birds inhabit those areas where boar have disturbed the soil. The dappled buttercup-coloured wood warbler, vanishing from many of our woodlands, fares significantly better in boar-managed French woods, preferring to nest on the ground in light vegetation. In addition, by digging small rodents from their dens, boar can also decrease predation pressure on such ground-nesting birds, displacing the small mammals that eat the wood warbler's eggs.

From the moment that boar turn drab uniformity into serrated chaos, new life moves in. Rich communities of ants often colonise these disrupted soils. These, in turn, can feed a range of predators. In the ancient woodlands of Białowieża, in eastern Poland, the wryneck – an anthill-raiding avian specialist sadly lost from our own shores – can often be found frequenting the anthills formed in the wake of rootling.

The frequent rainfall of our temperate island adds another layer of diversity to the wallows left by boar. Compared to our relict ancient woodlands, many of Britain's forest environments are extremely dry. Away from areas like the New Forest, where large tracts of deciduous woodland remain wetted by unfettered streams and bogs, many of Britain's plantations and woods are arid due to over-management and soil drainage. This means that amphibians, from great crested newts to our ever-declining common frogs and toads, must often travel enormous distances to find a suitable home across arid forest floors.

The wallowing of boar conspires with rainfall to change all of this. In areas of Britain being recolonised by boar, frogspawn can be found in the soggy heart of wallows; in other boar-ponds, all three species of newt. Boar wallows are shallow, creating, in turn, shallow ponds perfectly suited to amphibious needs. Boar have been known to eat frogs, as well, but it seems probable that when it comes to amphibians, the actions of far-roaming boar create more life than they destroy. Whilst boar systematically create habitat for amphibians, they are not known to systematically revisit every pond years later to eat those amphibians that have moved in. And whilst boar can, on occasion, consume reptiles such as smaller grass snakes and adders, they also create the perfect basking conditions for both species. Adders, in particular, will often warm up in areas of complex, disturbed soil, close to taller vegetation such as bracken, before heading off to hunt themselves.

Boar do not just leave unruly soils behind for others to do the work. They play a little-known but vital role in sowing the soil itself. *Zoochory* is a term that refers to the dispersal of diaspores: the seeds necessary to sow the next generation of plants. And wild boar, as they roam, aid *endo*zoochory: the dispersal of plant seeds through ingestion. Similar to their woodland neighbour, the red deer, boar can vector thousands of individuals of dozens of plant species around a landscape. Being habitat generalists, like red deer, and not tied to dense woodland, boar can also bring a range of grassland and non-woodland flowers into forest edges and glades. But unlike red deer, boar prepare the ground for planting as they go! Thus, those plant seeds fertilised then excreted by wild boar are far more likely to land in perfectly tilled gardens, receptive to their growth, than those defecated by non-digging mammals such as deer. In one elegant circle, therefore, boar prepare the ground and plant the flowers. Uncannily, we have only relatively recently learned to do this in our evolution as a species.

This, of course, resolves the little mystery at the start of our chapter. The many wildflowers, and their associated butterflies, adapted to the 'disturbed soils' now often banished to just our wastelands or roadside verges or the scruffier edges of some farmland fields, would once have prospered far more widely across all of our landscapes in the wake of wild boar.

Over time, the diggings of boar develop ever more complexity as trees begin to thrive. Wild crab-apple, an insect-rich haven for native wildlife, is beloved by boar. In autumn, boar, scrumping apples from the ground, are one of few animals capable of carrying their propagules intact within the gut. Roaming over large distances, boar effectively plant orchards, albeit those far more scattered and haphazard than our own. In Europe, a range of other now endangered trees, such as the Iberian pear, are most effectively dispersed and planted – complete with a healthy dose of fertiliser dung – by wild boar. And since the historical demise of Europe's elephants and giant elk, boar have become amongst

the important transporters – and planters – of large-seeded
wild fruit trees.

Where boar have dug in sunlit soils, in glades or grasslands,
sunlight-fuelled scrubland begins to grow, as earthworks are
rapidly populated with a rich variety of herbs and bushes. As
wild herbs, brambles and young hawthorns take root, new
creatures are brought into the woodland. In the twinkling
edges of forest glades, clusters of honeysuckle and bramble,
born from now-productive soil, become haunted by white
admirals. This butterfly, whose caterpillars feast only on
honeysuckle but which, in adult form, flickers between
bramble flowers in woodland, has recolonised many boar
diggings in the Forest of Dean, several years from their
inception. A range of now rare and endangered British
woodland butterflies, such as the wood white (which thrives
around the older boar diggings of the Dordogne's
woodlands), feed on a range of vetches and trefoils, which
grow best in disturbed soils.

One butterfly species in particular, the grizzled skipper,
has been studied in relation to the presence of boar. In the
Netherlands, it was shown that boar reduce the cover of
nectar-poor grasses through their shallow rootling, increasing
occupancy of the skipper's host plants. And such miniature
gardens, referred to by ecologists as 'pioneer microhabitats',
are virtually impossible to recreate, at a landscape scale, by
human hands.

Butterflies across Britain are in sharp decline as the
microhabitats they once used vanish from our ever more
sanitised countryside. In the absence of boar, conservationists
have been left to replicate a complex world of jostling plants
and met with limited success. Most butterflies of the British
woodland edge, once tied to complex processes like dam-
building beavers, wallowing cattle or rootling boar, continue
to vanish. Yet the boar's snout demonstrates remarkable
finesse when it comes to conservation. By transforming
huge areas of tall, species-poor tall grass to earth, boar sow

food-plants as they go. By crashing through fern, they bring sunlight to the forest floor, which fuels the growth of wildflowers. And, in 2010, the story of the boar and the butterflies took a twist that nobody had predicted.

In the heart of Sussex, the famous Knepp Wildland project, whose story is recounted in *Wilding*, by Isabella Tree, had restored free-roaming cattle, horses and, in its southernmost reaches, free-roaming pigs to the land. Wild boar cannot presently be legally introduced into the wild in Britain, so the Knepp project used, instead, a wild boar proxy: the Tamworth pig. The Tamworths were released into the rewilding project's southern block in 2009 – and left to snout it out. In the first years, their actions mimicked many of those described thus far for boar, as they rootled through Knepp's fields: transforming silent similarity into a chaos of new plants and trees. Yet, just one year after their introduction into the rewilding project, the finesse of a pig's snout would effect one of the most remarkable butterfly returns in British history.

Growing in the wake of rootling, but not in the un-dug soils in other areas of the estate, sallows had taken root. A form of hybrid willow, rarely tolerated in the farmed countryside or by many foresters, sallow is the sole food-plant of the purple emperor butterfly. This, in turn, lays its eggs on just a tiny proportion of individual sallows. For over a century, lepidopteran literature had turned the purple emperor into something of a legend, a canopy woodland butterfly, confined to the highest and oldest of our woods. Knepp's rootling pigs were about to unfurl a snouty surprise.

In 2010, the first purple emperor butterfly appeared, flashing through pig-aided sallows on gaudy, kaleidoscopic wings. There were no known colonies nearby, and the emperors were not expected guests. Since this time, Knepp's sallows have grown to host the largest colony of purple emperors in Britain, with more than 300 of these fierce, fruit-feasting giants on the wing each summer. In early July 2019,

I walked around Knepp to a riot of butterflies I had never
seen in my country; to a butterfly richness and abundance
that recalled my childhood forays in the boar-rootled glades
of the Dordogne. The arena had changed, but the emperors
and admirals, the skippers, whites and blues of the varied
scrubby grasslands were the same. And the architects –
Tamworth pigs, Knepp's stand-in boar substitutes – were the
snout-wielding butterfly gardeners behind this success.

Just as beavers get better in a landscape the longer they are
left to roam within it, the actions of boar – or free-roaming
pigs – in digging the land extend not only across landscapes
but also across time. As new scrublands grow in the wake of
rootled soils, a three-dimensional world of wild apartments
is created. In addition to the colonisation of Knepp by
purple emperors, equally remarkable has been the sharp
increase in turtle doves; a species now vanishing elsewhere
in Britain and headed for imminent extinction. For decades,
British conservation had pinned upon turtle doves the label
of a farmland bird; one that required the disturbed, weed-
rich margins of arable fields to survive. Like many species,
turtle doves had been making do – or failing to make do –
as herbicides slowly wiped from farmland soils the weeds
and seeds that turtle doves required to survive. Yet at Knepp,
as hawthorns, blackthorns and dog-rose pushed up through
snouted soil, an entirely new outcome was created. Here,
turtle doves are now nesting deep within a sanctuary of
thorny scrub, but feeding in the disturbed ground colonised
by plants such as chickweed, fumitory and knot-grass. The
rich thorny scrub grows in the wake of land rootled years
before. The plants grow in soils disturbed in the far more
recent past. In such scrub-grassland habitats, the action of
wild diggers are at once past and present. But in other areas
of Knepp, where the Tamworth pigs were not introduced,

the turtle dove has not prospected – its much-needed disturbed soil being absent.

Whilst a range of birds are adapted best to open grasslands, and others to dense walls of scrub, broken scrub-grasslands fuse these habitats into one and are amongst the most productive of all Britain's habitats on land. These habitats have also vanished over more than a century, taking with them a range of species whose futures now hang on the brink. The grey partridge feeds in disturbed soils but scuttles rapidly into cover; rarely straying far from thorny shade. Its nest is placed deep within vegetated scrub, yet its livelihood depends on foraging insects from bumpy open grounds. The vivid and cheerful linnet, increasingly absent from our farmlands, places its nest deep within bramble, but feeds around pioneer grasses and flowers. The cirl bunting, currently confined to the cavernous hedgerows and stubble fields of Devon and Cornwall, is adapted to nest in dense thorn, but forages on broad-leaved weed seeds such as fat-hen. The more we look to our 'farmland' birds, the more it becomes clear that these were once boar birds, too. The implications for their conservation, therefore, are enormous.

Were the actions of Britain's boar and their descendants to become more widespread in our grasslands, not only in our woodlands, then successes such as those being seen at Knepp might one day become considerably more commonplace in the regenerative farms of the future. And for this to happen, we have to remember that wild boar are, for some people, simply too *wild*. Although not dangerous to people, boar can disrupt productive farmed environments and cause financial damage on private land. Their widespread return to woodlands, wilder grasslands and even our wetlands could be welcome. But in other places, boar have a more widely accepted relative. Sadly, it too has now vanished from much of the countryside.

Nine thousand years ago – some time after we had
domesticated wolves, cattle, horses and sheep – settlers in
eastern Turkey had the wisdom and courage to domesticate
wild boar. Boar were most probably domesticated for a one
simple reason. Chasing them around woodlands with a spear
was dangerous and tiring work. Having boar on tap, in the
form of tame and well-fed pigs, led to fewer fatalities and
better outcomes for all – except the pig in question. Boar
remained wild, as wolves do to this day, but a new form of
farmland boar had been born. This brought about,
ecologically and agriculturally, a radical transformation.

For thousands of years, the pig would joyfully disrupt the
soils of farmed grassland Europe – and until the Second
World War, pigs were common across the mixed farms of our
own countryside as well; rootling at will. Many of the
onetime actions of boar would, in our farmland environments,
have been replicated, in part at least, by the pig. Old breeds in
particular, such as the Tamworth, the dappled Oxford Sandy
and Black, or the prehistoric and grumpy-looking Berkshire,
still travel and feed in a manner (when given the choice) that
is strongly reminiscent of boar.

In recent decades, pigs have become amongst the worst-
treated of Britain's livestock; penned indoors to a life of
misery that their good nature and intelligence does not
merit. They have become, in huge areas of farmed Britain,
almost forgotten as the farmyard animals that once roamed
free. Yet a return to wilder piggeries is not only in the
interests of a more sustainable farmed environment, and
better food, but also in the interests of our wildlife.

In particular, pigs can be profoundly beneficial when it
comes to restoring life to lifeless hillsides, compacted by
decades, if not centuries, of grazing sheep. Whilst sheep-
grazing results in soil compaction, and prevents regeneration,
pigs militate against this by rootling the ground and
maximising the chances of new flowers and trees taking
root, smashing through bracken and creating open ground.

On the regenerative farms of the future, pigs may have another role to play. As farmers begin to move away from deep-ploughing, which rips apart the fragile fungal structures of the soil, and weakens it over time, they might think to harness the oldest plough of all. By leaving old-breed pigs in a field over a length of time, you can rather effectively plough a field, albeit with less uniformity than a machine. The more pigs run free in our farms, on our hills, and in our woods, the more rich and diverse our lands will become.

Transforming the soil, resetting the basis of life itself, the boar and its friends must rootle back into our lands, our culture and our acceptance. As a nation of gardeners, all of us the friends of robins and the tillers of soil, we can surely grow to admire, and make some room for, one of the oldest and most talented gardeners of all.

CHAPTER TWO

Birds of Prey

Birds of prey are familiar to many of us as those imposing, regal and – for some – cruel creatures of the air, which prey upon land vertebrates for food. In the public imagination, species such as eagles loom large; symbols of wilderness and power. To others, birds of prey are simply a menace, a threat and a presence to be discouraged or, at least, despaired of, as they purportedly hunt down and destroy entire populations of smaller birds. In fact, whether seen as apex predators, or villains with talons, birds of prey are much misunderstood. One reason for this lies in the long history of ecocide that has taken place in our country.

Britain has for so long broken its contract with the natural world, and destroyed so much, so widely and for so long, that we have all forgotten the true numbers and diversity of birds of prey that would once have hunted our island – and the roles that they once played in enhancing the natural world around us. And even now, with many birds of prey recovering from the brink, we still enjoy these birds as a mere vestigial presence; museum relics of their former numbers.

If restored fully across the United Kingdom, our replete assemblage of birds of prey would have effects as surprising as they are profound upon the natural world. In fact, an abundance of avian predators within an ecosystem acts not only to regulate but also to enhance the lives of many other creatures, surprising as this may seem. And to investigate this further, we might begin not with the life history of one familiar bird of prey, the sparrowhawk – but the story of the bullfinch.

In the 1950s, bullfinches were everywhere. They were
hedgerow birds, orchard birds, field birds – and middle-of-
the-field birds. Not only were bullfinches blazing in the
hawthorns and the brambles, amid the hips and the haws,
but they lit up Britain's crops. Fruit growers were at their
wits' end as airborne shoals of secateurs sliced through their
profits. Wherever the bullfinch went, fruit buds vanished:
nibbled to oblivion by precise black bills. And so perished
whole crops of apples and cherries.

Bullfinches are ecologically designed to raid fruit trees.
Fruit trees are evolved to put out enough fruit to weather
this assault. What fruit trees are not evolved to cope with,
however, is a world where the bullfinch is in charge. And so,
for some years from the late 1950s until the 1970s, Britain's
fruit trees – invaluable sources of life for farmers, cider
makers and many wild birds – became imperilled organisms,
bereft of natural defence. It seemed that nothing could be
done. Yet within two decades, the bullfinch onslaught would
be driven into retreat.

The answer would come in the form of a low, silent
ambush. A swivelling bullet, slicing the hedgerow edge;
piercing the open gap. A projectile guided from the front,
steered from the back, the weaponry loaded underneath. At
the end of the 1970s, the sparrowhawk came back, as its
population slowly recovered from secondary poisoning by
the since-banned pesticide, DDT. And as it did, the bullfinch's
playing field was changed forever.

The true power of sparrowhawks does not lie in how
many birds they kill, but in the decisions they force other
birds to make. Just as a line of shops near a dangerous part of
some of the world's cities may empty of custom through the
mere *reputation* of its neighbours, so whole areas of habitat
become shunned as soon as a sparrowhawk reappears on the
scene. The landscape of fear is created as soon as one plump,
unwary bullfinch, fattened on fruits and light on vigilance, is
plucked in front of its field-dwelling peers. House sparrows,

the hawk's eponymous prey, favour areas with dense cover to dive into, beyond the reach of diving, grasping talons. In one study, it was found that redshanks tended to avoid certain areas of prime wet meadow as soon as sparrowhawks moved in. So fear-inducing is the long-tailed missile of the sparrowhawk's shape that other birds have copied it for effect. The female cuckoo, flushing grassland birds as she wickers from her low bushy perch, resembles a sparrowhawk. Panicked, pipits, dunnocks and warblers depart their nests, exposing the eggs to the cuckoo, among which she will, furtively, sneak one egg of her own. In short, sparrowhawks decide who feeds where.

The sparrowhawk begins nesting quite late in the spring – in the middle of May. Throughout a tense month, the male, who can occasionally be killed by the larger female, zealously brings his mate proteinous parcels of sparrows, tits, finches and other small birds. In the middle of June, the sparrowhawk's chicks hatch. The female, at this point, rises tenderly on muscled haunches, standing far enough above the chicks to allow them to breathe, yet shielding their sparsely downy bodies from the light of the sun. For the next few days, as the chicks' downy feathers sprout and flower, she will tend them closely. Then, with the chicks able to be left by themselves, all manner of chaos is brought to the woodland around, as both parents begin shopping in earnest.

A month earlier, birds such as blue and great tits have timed the hatching of their own chicks to coincide with the abundance of insects, especially aphids and caterpillars, found in the unfurling leaves of the oak. And now, young sparrowhawks hatch into the world ready to dine upon a veritable feast of dim, unwary and newly fledged insect-eating birds. Yet these apparently defenceless birds have prepared for the onslaught ahead. By laying between eight and ten eggs each, sometimes as many as 14, birds like blue tits future-proof their families; providing nature with a surplus of witless chicks.

June is the month that decides whether sparrowhawk chicks live or die. Each day, up to a dozen small, de-feathered meatballs will be brought to a sparrowhawk's nest. Brutal as the onslaught may seem to human eyes, these predators are merely harvesting the surplus. And in doing so, the role of the sparrowhawk, whilst not protective of the blue tit, becomes protective of the wider woodland environment in which it lives. But how?

Hunting aphids by the thousand, blue tits are the pest-removal service of our woodlands, ridding us of more plant parasites each year than all our gardeners combined. Yet aphids, too, play a fundamental role in the health of our woodlands. Removing too many aphids would have far-reaching consequences, most of all for our most industrious of engineers – the ants. And this is what would happen, if blue tits were left unchecked. In fact, the ramifications of blue tits having life all their own way could be profound.

Let's take the southern wood ant – a crucial species in the diet of green woodpeckers. It will commonly climb 30 metres into old oaks, birches and pines, where it will begin a farming operation perfected over millions of years. Gently stroking an aphid's abdomen, the wood ants elicit the production of sucrose-laden honeydew. Those aphids stroked by ants, it has been found, tend to produce smaller droplets of dew, richer in the amino acids that ant societies need to grow their young. In turn, the ants will zealously guard 'their' aphids from attack, spraying formic acid at any predator that dares to intrude. Then, their abdomens warped and bloated with rich sugars, the ants scale back down the furrowed bark, bringing honeydew to their queen and her workers, and regurgitating it to feed the brood. In doing so, species like the pine aphid fuel the formation of Britain's most spectacular ant cities. And growing to a metre in height, wood ant nests provide a bounty of food for other birds. What's more, wood ants also act as predators of defoliating insects, protecting our native trees from harm.

Ants, by moving soil and collecting insect food, alter the composition of our soils. Ant nests, in particular, have been found to dramatically improve the abundance and diversity of grassland plants, and insects, within the vicinity of a nest. Too many ants, however, reverse this effect. These powerful micro-predators, fuelled by aphids, are kept in check by the actions of aphid hunters such as blue tits. But for such nuanced societies to survive, the role of sparrowhawks in harvesting the aphids' hunters is even *more* important.

Perhaps, given any thought, the destruction of a delicate treetop aphid farm by a blue tit is no lesser form of natural vandalism than the sparrowhawk plucking a blue tit on your garden lawn. Yet, in spite of their important role in protecting aphids, ants and grassland diversity, the role of the sparrowhawk is much misunderstood. Even now, pseudo-scientific charities in Britain have looked to the decimation of our insectivorous birds in recent decades – and picked a villain on whom to pin the blame. Ignoring food loss and the loss of large areas of sympathetic habitat, it has been easy to demonise the sparrowhawk.

Yet it is not in the nature of a sparrowhawk, nor in its fundamental interest, to wipe out the prey that it depends upon for survival. In the longest-running study of any bird of prey, the renowned professor Ian Newton found no evidence that sparrowhawks wiped out the animals beneath them on the trophic ladder. Rather, with sparrowhawks absent, rather than dying of predation in summer, prey populations such as blue tits were limited, instead, by a slower starvation across the winter months. By this time, however, the damage wrought by a surplus of blue tits upon insect communities would already have been done. In terms of ecosystem benefit, then, it is better for young birds to die in summer than to overburden our woodlands, and their resources, come the autumn.

In reality, the role of the sparrowhawk is far more remarkable and beneficial than the simple killing of small

birds. Whether protecting aphid-reliant ant societies or
fruit trees, sparrowhawks – both through the predations
they carry out and the fear-factor they engender that
changes the behavioural patterns of birds within their
territory – act to protect and enhance the lives of both
insect and plant communities. Yet, for all its yellow-eyed
ferocity, the sparrowhawk, too, is designed to be served for
dinner. In the ocean, it is said that there is always a bigger
fish, and in the taloned world of birds of prey, the very
same rule applies. In fact, the sparrowhawk is not, quite,
where the food chain ends.

A female goshawk, bulging with a kilogram of muscle,
five times the weight of a female sparrowhawk, is a
formidable predator. For 10km² or more around her
home, a host of smaller taloned creatures must hide, flee
or die. Goshawks are the hunters both of adult birds of
prey but also, more often, their young. They do so for
food – and also to reduce competition. This rarely seen
yet normal behaviour, known as 'intraguild predation', is
one means by which smaller aerial predators are kept
in check.

At the time when 60 per cent of Britain was covered in
large tracts of Atlantic rainforest and vast wood pastures, it is
thought that as many as 20,000 pairs of goshawks may have
hunted our island home. Their dominant presence in all
largely wooded areas would have ensured that sparrowhawks
and kestrels, and even buzzards and kites, remained refugees
in those of our landscapes – such as scrub grasslands or river
valleys – less suitable for the goshawk's needs.

Eradicated by the end of the nineteenth century, the
goshawk's loss has imbalanced the countryside for centuries.
And whilst it has now returned in good numbers to Wales,
western England, the New Forest and areas of Scotland, we

still enjoy goshawks as prized rarities, rather than the dominant woodland predator of our wider countryside.

Goshawks require large tracts of woodland in which to hunt, though they hunt it with enormous flexibility. From the dark plantations of upland Wales to the broken deciduous woodlands of Herefordshire, and the ancient wood pastures of the New Forest, the goshawk thrives and kills in them all. Given a complete absence of persecution, as seen in Germany, for example, so adaptable is this woodland hawk that it can hunt as effectively in Berlin's city parks as in the larches of the Forest of Dean.

Goshawks will rip kestrels from their nest without a second thought, and pummel young buzzards and honey-buzzards to death. Video footage has showed Polish goshawks calmly flying off with the large chicks of lesser spotted eagles. Recent nest-cam footage from the BBC's *Springwatch* even showed a pair of goshawks pelting towards an osprey eyrie. As the larger osprey takes off to tackle one adult goshawk, the other deftly flies off with the osprey's large chick, with barely a second's delay.

Sparrowhawks are also high up on the taloned menu for goshawks, and frequently found, plucked, below goshawk nests. Young tawny owls, unwarily 'branching' from their tree-cavity nests, are transformed from downy to mere down. A host of other species, including grassland-nesting short-eared owls and copse-nesting long-eared owls, can be removed from the territory of goshawks, as was studied at some length, again by Professor Ian Newton.

In the 1980s, the goshawk returned to Northumberland's Kielder Forest and to the Scottish Borders. A large area of dense spruce plantation intersected with rough, vole-wriggling moors, Kielder was home to a wide range of common birds of prey. Kestrels hunted small birds and rodents at its margins. Ground-nesting short-eared owls floated over its grasslands, seeking the field voles that constitute almost all of their diet. Buzzards, recovering well in most of their range as rabbit populations increased, were abundant. The goshawk's arrival transformed the state of play.

At one goshawk nest in the Borders alone, 20 newly plucked kestrels were found at the end of summer. Just as sparrowhawks time the hatching of their young to coincide with an abundance of the young and unwary, so goshawk chicks demand the most protein during the time when young raptors of other species are also in the nest. As well as eradicating competition, the killing of smaller predators by goshawks feeds to their own chicks a prime steak's worth of meat, the muscle richer and denser than that of a woodpigeon or squirrel. Within a decade, almost all kestrels, and most breeding short-eared owls, had vanished from Kielder Forest.

Whilst an onslaught like this may seem ferocious and destructive, such an apparent killing spree is generally what happens when a cornerstone predator, absent for decades if not centuries from much of the landscape, makes a sudden return. Just as the wolves in Yellowstone killed more than 20 elk per wolf pack member in the years after their reintroduction, and wiped out a large number of coyotes, so British goshawks have got to work in rebalancing our own long-imbalanced aerial food chain. But as they have done so, remorselessly ripping out the smaller predators, the beneficial effects have been profound and surprising.

Goshawks are not only the predators of other birds of prey. In addition to squirrels, woodpigeons and a range of larger prey (including ducks, and even birds as large as greylag geese), it is the predation of crows that renders goshawks so important to the countryside. Prior to the goshawks' return to Kielder, huge marauding gangs of non-breeding carrion crows were wreaking havoc. Barely ever hunted by sparrowhawks, there is nothing more devastating to landscape diversity than an army of crows with time on their hands.

Carrion crows, as gamekeepers will correctly warn, are a threat to the life of wading birds. Sitting at leisure upon a tree, bush or telegraph post, crows will watch ground-nesting birds, like curlews and lapwings, as they return to their nests and will then make off with all the eggs, or as many newly hatched chicks as they can capture in the grass. In doing so, carrion crows, as well as magpies and ravens, can dramatically reduce wader breeding success.

The presence of goshawks changes such a dynamic. Indeed, on returning to a breeding territory, one of the first game-keeping services provided by goshawks is the removal of a large proportion of a landscape's magpies. Clever at planning but terrible at flying, a magpie is no match for the aerial agility of a goshawk. Carrion crows are often next to go – since the goshawk's return to Kielder, huge reductions have been seen in the crow population. Furthermore, jays (which raid the nests of many other species) must become ever more wary if they are to survive. In controlling the predators of nests, goshawks, which seldom bother with protein parcels as small as eggs or newborn chicks, can therefore increase the chances of survival for some of our most vulnerable birds.

In the rich woodlands of the New Forest, one of many birds thriving here, yet vanishing elsewhere, is the enigmatic hawfinch. A powerful parrot-like finch whose black eye-mask gives it a perennially angry expression, the hawfinch has a seed-crushing bill that can slice through cherry kernels or, in the case of one unfortunate bird ringer, the bones of a human finger. Hawfinches are canaries in the mine when it comes to the health of a woodland. Not only do they need large tracts of deciduous woodland to survive, but they also favour the oldest and most diverse of woods. Oaks provide them with moth caterpillars to feed their chicks; hornbeam and beech provide them with fallen mast in the autumn months; and crab-apples and cherries, rare in many newer planted woodlands, can be vital sources of food come the winter. In the past three decades, hawfinches have vanished from most of our woodlands, including almost all of the smaller ones. This has happened for a number of reasons, including the loss of wood-pasture habitats and older deciduous trees within the landscape, the loss of a rich diversity of trees from many habitats – and also, nest predation.

Hawfinches now persist, in good populations, in just five areas of Britain, and in many of these they continue to prosper. From the woods of the southern Lake District to north-west Wales, the Forest of Dean and Wye, and the New Forest, the hawfinch's last strongholds share two things in common – large tracts of mature deciduous trees, and, more surprisingly, a good supply of goshawks. Why?

In spite of their fierce, formidable appearance, hawfinches are noted for being birds that build surprisingly flimsy nests. Many nests, like those in wild apples, are simply placed on branches, where the fluffy chicks stand out like pompoms on a tree. Or a nest might be placed in the ivy-clad cleft of an oak tree – often the first place that an aerial predator such as a jay will alight. Even in well-regulated woodland ecosystems like Poland's Białowieża Forest, less than 30 per cent of hawfinch nests survive predation. Yet even this number is enough to ensure their survival. In other words, hawfinches have always been predated at the nest, and most nests fail. But their nest predators are supposed to be regulated, too. Jays are the main predator of hawfinch nests – and this is where the goshawk comes in.

Fieldworkers in the New Forest have found that hawfinches cluster almost all of their roosts, and many of their nest sites, within 200m of an active goshawk nest. Nest raiders like magpies and carrion crows, which can haunt the woodland edge, are often removed from play. Jays may still be present, but in lower numbers. In addition, the same fear factor that discourages bullfinches from open fields means that jays dare not linger long in the heart of a goshawk's territory, reducing their time to find and predate hawfinch nests. Goshawks, by contrast, will rarely expend significant energy chasing birds like hawfinches, or other small birds, for food.

In helping to protect hawfinch roosts, goshawks allow these strange, parrot-like birds to carry out a range of woodland services. Hawfinches are one of few species capable of tackling the oak processionary moth, a sinister danger to oak trees if left uneaten. Most of all, they are the only aerial disperser of the wild cherry: no other flying species can penetrate its kernels.

In turn, by muscling buzzards and sparrowhawks out of a landscape, and eliminating a large proportion of its crows, goshawks are one of the most powerful and effective of winged gamekeepers, creating habitats where endangered smaller birds, and wader chicks, are more likely to survive. Yet, as the studies in Kielder Forest have shown, a goshawk unchecked can remove not some but *all* of the smaller birds of prey in a woodland. And if rodent-hunting species like kestrels vanish *completely*, there are other, less-welcome ramifications. Rats, for example, would greatly increase – and that, in turn, might lead to increased predation of the eggs of ground-nesting birds! As with anything involving birds of prey, all is not straightforward. Nature does, however, have a trump card designed to handle an ecosystem surfeit of goshawks. It's giant, it's silent, and a goshawk never sees it coming.

In parts of Europe such as Fennoscandia, a strange distribution befalls the owl population. Often, you will find a landscape full of middle-sized owls, such as tawny owls – but few small owls. In other landscapes, you will discover very big owls, and very small owls – but hardly any middle-sized owls. This peculiar situation arises because owls, like diurnal birds of prey, specialise in eating one another as well.

Both Ural and eagle owls, for example, will actively prey upon tawny owls, but seldom bother with the irate, feathered tennis ball that constitutes the pygmy owl. Pygmy owls, however, are often eaten by tawny owls. When giant owls are present in a landscape, however, they create the space for the smallest predators to survive. This came to the attention of ecologists in Germany, who were faced with a dilemma – the return of the goshawk was welcome, but their numbers had grown exponentially, and impacts were being noted on other species, such as honey-buzzards, at a landscape scale. So the scientists turned to the goshawk's only natural predator – the eagle owl.

Of all the birds in Europe, nothing comes close to the predator clean-up operation run by eagle owls. In Norway, a third of all birds eaten by eagle owls are other birds of prey. Long-eared and tawny owls, buzzards and kestrels are commonplace prey – but so is larger game: goshawks, peregrines, and even the relatively large chicks of ospreys and white-tailed eagles. Since the reintroduction of eagle owls to Germany's forests, densities of goshawks have declined by up to 50 per cent.

The secret to these extraordinary assassinations is that whilst those other raptors fly during the day, eagle owls hunt at night, when the raptors are tucked up and asleep. What would be an internecine conflict of talons, bills and blood by day thus becomes a swift and silent death by night. One nest camera, mounted on the cliff-side nest of a long-legged buzzard in Israel, gives some idea of how little chance a daytime raptor stands against a nocturnal super-sized owl. In the video, two young buzzards are dozing. Then, a fox-sized bolt from the black pelts out of the darkness, seizes one of the buzzards, the same size as itself, and vanishes into the gloom without even landing. The

entire event takes two seconds. This is how eagle owls prey
not only on smaller owls but also on goshawks, ospreys and
even young eaglets. In some cases, the eagle owl will even
eliminate *all* the birds of prey within a territory. As such, its
effects on the ecosystem are profound. Whole orders of
small day-flying birds, susceptible to daytime predators but
secure in well-hidden nests by night, benefit greatly from
the presence of eagle owls.

In the history of British birds, none has a more mysterious
record of occupation than the eagle owl. We know these
were native birds; fossils prove their existence as firmly as
they do that of our elephants and aurochs. Eagle owls were
present around Demon's Dale, in Derbyshire, 10,000 years
ago, but after that, the fossil record goes quiet. Eagle owls
may have vanished from Britain even before the aurochs
and Dalmatian pelicans. In some ways, this might not come
as a surprise.

The eagle owl might not have been the friendliest of
neighbours. Professor Ian Newton points out that not only
does the call of the eagle owl, a haunting 'boom, boom' that
bounces around inside your skull, and shivers your spine,
carry for miles (and would have engendered much fear and
superstition), but eagle owls are also amongst the fiercest of
birds when defending their chicks. Their favoured rocky
bluffs, bulging from sheltered hillsides, would have been the
perfect homes for *Homo sapiens*. We know from cave
paintings in France that humans watched cave lions and
their prey from the safety of such vantage points. Sharing
your home with a giant owl may have led to a conflict that
could only end one way.

So incomplete is our fossil record of eagle owls, rather
like the bison that must once have graced our shores, that
tracing their departure proves difficult indeed. The discovery
of a few bones hint at their presence in Somerset until 2,000
years ago, whilst reliable Victorian naturalists were describing
eagle-sized owls in Scotland considerably later. For now,

eagle owls rest under the label of 'non-native' species; a strange one indeed, given that they nest from the trees of Finland to the rock faces of Calais, within sight of Dover's white cliffs.

Whilst eagle owls are long forgotten as native predators, and goshawks only beginning the process of attaining the numbers needed to influence ecosystems here in Britain, the absence of these predator-killers from large areas of the British countryside has led to other predators, those designed for the 'middle' of the food chain, to come out on top. And that brings us, in turn, to the extraordinary success story of the common buzzard.

The aerial fox of the countryside, mobbed by everything from crows to starlings, it can be easy to mistake the buzzard for the bumbling buffoon of the avian predatory kingdom. Lacking the muscular lethality of a goshawk or the pace of a peregrine, these lumpish raptors can neither soar as well as an eagle nor hover as well as a kestrel. But watch a buzzard for long enough, and you will witness a dozy brown lump transform into a rabbit-killing torpedo. Buzzards, when they wish, can stoop at over 100km per hour, and aerially ambush a range of wary prey from woodland jays to fleet-footed squirrels. A surprisingly large part of a buzzard's diet comes from earthworms, but they will also harvest many of the lower levels of the food chain, from nestling and newly fledged birds to voles and amphibians. Buzzards share certain tendencies with their fiercer woodland cousin, the goshawk. They will frequently skim kestrels from their nests and muscle these small falcons off nesting sites in quarries and on cliffs. At the same time, buzzards can afford powerful protection to the small. By preying on large quantities of rodents, which can eat the eggs of ground-nesting birds, buzzards act to protect species such as meadow pipits, skylarks or corn buntings.

As buzzards have bumbled back into British airways, they have been greeted with applause by nature lovers. Now they grace the skies of every county in our island. In the wave of enthusiasm, conservation groups have argued that buzzards, as well as red kites, are welcome 'apex predators'. But in the red-taloned world of raptors, buzzards do not rank highly: classic mid-order or 'meso' predators, they are more suited to a role of limited power. If one reason for this is the power of the goshawk within woodlands, the other is merely a matter of size. Buzzards, in turn, would once have bowed to eagles.

In many ecosystems, from the river valleys of southern Africa to the Spanish hills, from the lakes of northern Finland to the coast of Japan, all functional ecosystems need giant aerial scavengers. As far as we know, Britain has never played home to vultures, but nor has it needed to do so. Britain's vulture niche was, and is now again being filled by a far more versatile bird – the white-tailed eagle.

Prior to the Bronze Age, as wild horses perished from old age in winter hilltop blizzards, the white-tailed eagle would have closely followed ravens to a carcass, scattering them as it alighted. As the written records of Saxon battles attest, as the dead fell, so eagles would swoop from the skies. Yet the white-tailed eagle has a versatility rare in any bird of prey. Not only a consummate scavenger, it can also kill a greylag goose on land as effectively as it can rip a salmon from the water. Whilst tied largely to water, it can nest anywhere from giant pine trees to hanging willows and, if short of options, it will even build floating platforms in the reeds. Hunting deltas, floodplains, ponds and the coast with equal ease, white-tailed eagles are a species built for versatility, abundance – and *dominance.*

So persecuted have eagles been over centuries, so degraded their habitat, that we have come to cherish them as rare. Yet the designs of an ecosystem necessitate that white-tailed eagles, like any apex scavenger, should in fact be common. And when undiminished by habitat loss or human depredation, they are. Protected since 1919, Russia's Volga Delta is one of the only places in Europe where the preservation of original habitats has been matched by a century's fierce legal protection. Recalling the Somerset marshlands of 3,000 years ago, the Volga reminds us of a world that only Britain's Neolithic ancestors would remember. In all, one *thousand* white-tailed eagles feed and nest here in the Volga, in an area of willowy marshland smaller than Norfolk.

Extrapolating from such natural densities, contemporary estimates that Britain held around 1,400 pairs of white-tailed eagle, prior to their eradication, seem extraordinarily low. In the Bronze Age, large wetland systems covered 20 per cent of Britain – the greatest being in the mosslands of Cheshire, Somerset, the Humber estuary and the huge fens of present-day Lincolnshire and Cambridgeshire. Each of these wetlands alone may have been large, and prey-rich, enough to host 1,000 white-tailed eagles.

Early on, however, Mesolithic settlers moved into these marshlands, living on boardwalks and in small boats. The audacious white-tailed eagle would have competed with them for food. Centuries later, it is thought medieval monks, tending fish ponds rich in carp, would also have been 'plagued' by these birds. As we drained Britain in the seventeenth century, we would drain our lowland eagles. For centuries, we drove them to our coasts, then wiped them out. Birds remained for longer in Scotland – and eventually we cleansed them from here as well. Yet even as the white-tailed eagle was being hunted out of Britain in the early nineteenth century, we are reminded of its onetime abundance here on our own shores.

The records compiled in Roger Lovegrove's *Silent Fields* reminds us that on one Sutherland estate alone, home to numerous fish-rich lochs, 171 adult eagles and 53 young were wiped out during the 1820s alone. Such localised examples of abundance suggest that centuries before, the lowlands, sea bays, wetlands and rivers of Britain may have been home to several thousand pairs of white-tailed eagles. These birds had evolved, like the vultures of Africa and India, in an age of plentiful herbivore carrion, and fish. They had evolved to be common.

In recent decades, thanks to pioneering reintroduction schemes, Britain's 'flying barn door' has recolonised a little of its former range, with more than 100 breeding pairs. At the time of writing, birds reintroduced to the Isle of Wight are now gracing English skies for the first time in centuries. Yet today, white-tailed eagles remain so rare that we have forgotten the fundamental role their circling flocks once played in altering the landscape. Their effects upon fish, and fishing, would have been profound.

As well as eating their fair share of nestling buzzards, and displacing others from their territories (buzzards would never have attained their numbers here in Britain, had the white-tailed eagle been present in its former numbers), white-tailed eagles bring to heel that most 'insatiate' of birds, the cormorant. In northern Germany, entire colonies are forced away from prime fishing grounds due to heavy predation on their chicks. Cormorant colonies coming under attack are prone to panic; fleeing their guano-washed nests, they leave their chicks to be harvested on repeated visits by eagles.

Heronries, too, are often moved or displaced by the threat of white-tailed eagles. Large colonies, capable of reducing fish productivity, are fractured into smaller ones. Scattered across the landscape, these refugees come to have less effect upon a wetland, dispersing the prey-base for a wider array of other species to profit from. Birds like herons and cormorants are elegant and graceful, yet the latter, in particular, is one of the most intensive hunters on the planet. Unchecked, cormorants can decimate fish-stocks; the white-tailed eagle prevents this. In suppressing the most demanding anglers in a wetland, yet taking only far smaller quantities of fish themselves, white-tailed eagles ensure that sufficient stock exists to replenish the population the following year.

In undiminished abundance, the effect of gregarious fish-eating eagles grows greater. In Alaska, the bald eagle has recovered to a far greater degree than its European cousin. Each autumn, thousands of eagles, dispersing from breeding grounds along the coast of British Columbia, gather to feast on the post-spawning carcasses of fish, in rivers that still run red with salmon each fall. In scavenging the salmon, flying with them to perches and losing others during aerial squabbles, bald eagles fulfil the aerial role of bears, dropping the remains of nutrient-rich salmon to fertilise the land. Long ago, when British coasts wriggled with spawning salmon, white-tailed eagles would have carried these ashore. In doing so, these birds would have played an important role in fertilising the lowland woodlands of our coast – acting as nutrient vectors, transforming the very composition of the soil – and the trees that grew within the Atlantic rainforest.

White-tailed eagles may be the largest birds of prey ever to have graced our island, but alighting at a carcass, even they can defer to one other. Peregrines flee before this eagle. Buzzards and hen harriers move home to avoid it. It regulates the land with some of the keenest eyes on Earth and talons that can pierce the backbone of a wolf. Of all our flying creatures, not since the age of dinosaurs has one more powerful graced Britain's skies.

Golden eagles are regarded the world over with reverence and awe. They are the most powerful aerial hunter in much of the world, outside the tropics. With a grip 15 times the strength of ours, they know few limits in the prey they are able to capture. Documentary video evidence shows that golden eagles trained by Mongolian falconers can not only capture but also dispatch adult wolves, though fights do not always go their way. In Britain, livestock predation is extremely rare, yet in one extreme but well-documented incident in New Mexico, a particular population of eagles felled 12 domestic cattle calves, some exceeding 100kg in weight. Pairs of golden eagles in North Macedonia have been known to specialise in horses.

The scale and audacity of the golden eagle's hunting operation is known only to a few. In Sweden, a young lynx was found in a nest. Young otters and badgers have been brought to Scottish eyries. Red foxes are commonly taken, making up more than a tenth of the eagle's diet in places such as Sicily. Roe deer are not only frequently killed but also carried to the nest: in the Italian Alps, up to a third of the nestling diet consists of this fleet-footed ungulate. Red deer calves can be carried off. On the Swedish isle of Gotland, half of the eagle's diet consists of hedgehogs. In Alaska, brown bear cubs have been whisked away. In Europe, Hermann's tortoises have gained surprising turns of speed in their final minutes. Across mainland Europe, wild boar piglets have gained the power of flight.

By contrast, records of golden eagles being killed are as rare as might be imagined. In one case, an overambitious bird died in a fight with a snow leopard. In Scotland, an eagle seeking to attack wildcat kittens died in a fight with their irate mother, who was also killed. In Alaska, only adult brown bears regularly dare to raid the ground nests for eaglets to eat. Barring perhaps the orca, golden eagles may be the most widely feared of the world's apex predators.

Whilst small birds sharing the realm of eagles have little to fear, the same cannot be said of the birds of prey, crows or other large birds that fly where eagles dare. In Estonia, cranes are regularly eaten, and, in northern Norway, mute swans. Golden eagles will regularly kill buzzards and hen harriers, or push them into areas of marginal habitat. They will drive peregrine falcons from their nesting cliffs. In the Galloway hills, in the 1940s, when golden eagles returned after a century of persecution, each pair would displace a peregrine from its ancestral cliff-top home. In Europe, even the Ural owl – a huge, deceptively calm-looking owl famous for striking at the eyes of humans – is not beyond a golden eagle's reach. Even goshawk-killing eagle owls have, on occasion, been discovered in an eyrie. In most direct interactions with white-tailed eagles, golden eagles win. At carcasses in Finland, the larger eagle inevitably gives way to the superior talons and supercharged ferocity of its muscle-bound predatory peer.

So early did golden eagles lose their wilderness and habitat in Britain, we have all but forgotten them except as Scottish birds. For centuries, no one has watched golden eagles harry deer on the hills of Snowdonia, or power through Dartmoor's steep valleys in search of woodland grouse. Yet these sights were once our ancestors' to enjoy.

Reconstructions of the golden eagle's former range in Britain have found it present throughout the hills of northern England and Wales. The fossil record records eagles from Fox Hole Cave in Derbyshire, an area of low hills east of Chesterfield. Their bones have also been found at Catterick, an area of moorland in Yorkshire, and in the Ossom's Eyrie Cave in central Staffordshire. None of these sites, incidentally, falls within the high, forbidding mountains that we might associate with this species in Britain today.

Not only do bone remains from Stafford Castle date into the sixteenth century, but reliable eyewitness accounts are

also surprisingly recent. Eyries were found in Derbyshire's
Derwent Valley until at least 1668, and in North Yorkshire
until the 1790s. In Wales, eagles were still nesting on the
cliffs of Llangollen in 1656, and until 1800 in Snowdonia.
More recent studies have further 'expanded' the range of
Britain's aerial masters. One study, analysing old place names
in relation to the habitat of white-tailed and golden eagles,
has found good evidence that golden eagles hunted
Dartmoor a millennium ago, whilst some naturalists'
accounts appear to reliably detail golden eagles on Dartmoor
as recently as the early 1800s. Yet even these records are
unlikely to reveal the whole picture.

As with so much of what we have lost, 'shifting baseline'
syndrome has led to ecological amnesia amongst many of
Britain's naturalists. We have assigned golden eagles to steep
crags and mountain ranges, forgetting that they have, over
millennia, become mountain refugees. Studies have assumed
that eagles stop where mountains end. Yet given any choice
in the wilderness they hunt, golden eagles can breed and
thrive in lowland areas as well.

In countries like Estonia, where the undisturbed lowland
forests lost to Britain in the Bronze Age still grow intact,
eagles build their largest nests in giant pines. Here, they
breed successfully in remote areas where grouse, deer and
hares form much of their diet. Estonia's low open wetlands
are ruled by white-tailed eagles; its remote, broken forests by
their golden cousins.

In addition to enormous areas of dry forest – far from
water and unsuitable for white-tailed eagles to hunt – rich
lowland forest and bogs would have been found within
many British wetlands. The fossil record details golden eagle
bones from Meare Lake, in Somerset, as late as the Iron Age.
If not from Somerset's marshlands itself, perhaps this eagle
had ventured over from the Mendip Hills, where its bones
have been unearthed in Cheddar Gorge. Dense Atlantic
rainforests, greening the steep, cliff-rich hills of Wales,

Exmoor and the Mendips, would, prior to their deforestation, have suited golden eagles to perfection. Now, long forgotten by us, the British landscape still feels the gaping loss of its golden eagles.

In regulating all flying predators, deer and ground predators such as foxes, golden eagles free up whole orders of the animal kingdom to go about their daily business, and can even protect ecosystems against overgrazing. By pushing small hunters to the margins, golden eagles guard the realm of the small, the fragile and the young. In preying on young foxes and badgers, golden eagles create a landscape of fear, where ground hunters dare not wreak havoc for long. And in taking such a broad diet of animals, as generalist predators, golden eagles never exert too much pressure on the survival of any one species. Their role in the wild is as finely balanced as their perfect silhouette. And the golden eagle is the final, greatest piece in the seemingly paradoxical puzzle of birds of prey: those species who, through killing, simultaneously protect and enhance the natural world.

It is very hard, however, for many British people, or indeed those in most of the industrialised world, to grasp the true power and importance of birds of prey – large and small – for the simple reason that we now have so few. But what would happen if these diverse hunters were restored to their original numbers, across ever-expanding tracts of suitable habitat? What would happen if we briefly turned back the clock 5,000 years, to a time when eagles governed every British airspace? To get some idea, we must look, for now, to those parts of the world where birds of prey have never vanished at all.

In 2019, I spent over eight weeks in the Luangwa Valley of Zambia, an area of untouched river valley and wood-studded savannah larger than Yellowstone National Park. Few places on Earth may be considered truly pristine – but Luangwa is as close as it comes. For centuries, sleeping sickness and the impossibility of farming has preserved an ecosystem not dissimilar in magnificence to that last seen in Europe during the Pleistocene Era. Whilst the ever-vying lions, wild dogs, leopards, hyenas and crocodiles are the familiar ground predators that help protect the diversity of this landscape – preying upon the sick and weak, and preventing Luangwa's diverse herbivores from grazing down its scrubland, grassland and trees – fewer tourists here look to the skies. In reality, one of the greatest guarantors of biodiversity in this landscape is a teeming abundance of different birds of prey.

South Luangwa National Park, covering 9,000km^2 of Africa, has seen the poaching of rhinos in recent years, but there is no evidence that its birds of prey have ever been persecuted, or their habitats fundamentally altered, in the past 10,000 years or more. Everywhere you travel, birds of prey abound, each governing its niche within a hostile wider kingdom. Large kettles of vultures, in their dozens, forever scour the landscape, as white-tailed eagles once did in Britain, seeking fallen herbivores; some felled by disease, others by lions, hyenas, leopards or dogs.

Martial eagles, capable of lifting impala from the ground, relentlessly hunt grazing animals from gazelles to scrub-hares; protecting, as they do, the grasslands and new trees, none of which would form were small herbivores alone left in charge. Lizards abound in the grasslands, scoffing many large grasshoppers on which shrikes depend – but many will be carried off by the lizard buzzard. A rich chorus of scrub-dwelling songbirds, denser than any left in Europe, haunts Luangwa's diverse bushlands, but they will never get the chance to harvest more than a portion of its small invertebrates: each is being watched by predators from the little sparrowhawk to the Gabar goshawk – no sooner has prey filled a niche than a predator is on the scene. Fish flourish across the river valley, but they will not divest it of small, aquatic life; a large population of African fish-eagles see to that. The plentiful African guineafowl here feast upon beetles and weeds, but they will decimate neither: the African hawk-eagle keeps them in check. Venomous snakes haunt the grasslands here, killing many of the rodents that would eat birds' eggs. But they will never eat them all: no fewer than three species of snake-eagle act as airborne anti-venom. Ground-nesting waders, such as lapwings and thick-knees, may be sometimes plagued by birds that rob their nests. But should those robbers hesitate an instant, a hawk, buzzard, falcon or eagle would be upon the scene. In Luangwa's skies, for every check, there is a balance. For every order in the animal kingdom, forever threatening to grow out of control, there is a bird of prey. One day, we may once more enjoy the same here in Britain.

Even now, it is possible to restore birds of prey to all corners of our land. And as we restore our ecosystems, we will begin to see the effects of birds of prey once again. From the sparrowhawk's safeguarding of wood ants and fruit trees to the goshawk's protection of the hawfinch; from the white-tailed eagle's protection of fish stocks to the golden eagle's dominance over buzzards and crows, all of our aerial predators have a vital role to play – the protectors of a song-filled world that we all wish to inherit and enjoy.

CHAPTER THREE

Beavers

Weaving its way through the gritstone moors of the Peak District runs the Bar Brook. Welling in the swampy Totley Moss, it drains large parts of Big Moor as it snakes into Barbrook Reservoir, then tumbles off the moorland edge. Gaining speed, the brook becomes a torrent as it pelts through a steep ravine to join the great Derwent river. From the air, Bar Brook, for most of its course, looks as if it's been folded several times, like unruly string in a drawer. In a straight line, the distance covered by the Bar Brook is just 8km. Bent double with winding sinuosities, it runs twice that distance.

The moors scoured by Bar Brook are rich in relics. On the river's northern bank, an old smelt mill, closed by 1770, lies in ruins. Bronze-Age cairnfields, well preserved, attest to the fact that more settlers once lived here than today. Snaking through the ghosts of Neolithic farms and Georgian mills, the Bar Brook travels through the crumbling remains of lost function. The cairns' true purpose, the mill's utility – all have all been forgotten over time. Only one monument here has defied the erosion of the years – and of the water.

Even in full spate, the brook's cascading flow has been altered forever. An ancient system of ponds slows its every move. At each turn, a myriad of miniature enclosures siphon the speeding torrent, calming its intent. These impoundments were designed to slow and change the water's course. Restraining the brook, frustrating its pace, they have stood firm for centuries. Unlike the long-eroded mills, both their form and their function have survived.

Forgotten by anyone alive, the complex society that reformed this moorland landscape has long since departed. In fact, beavers, known to the local people of the time as 'bars', haven't called the Bar Brook home for more than 700 years. Yet even now, their impact on the landscape, and its water, stands the test of time.

Beavers transform our water worlds and the lives of those who dwell there. They are the cornerstone species of freshwater systems. Beavers manage wetlands better than our finest conservationists. They conspire with time to create the most diverse of landscapes, and slow the powerful flow of fast-running rivers in more innovative ways than the most skilled of human engineers. Yet the irony of beavers lies in a profound ignorance of their own importance. What beavers do, and the busyness with which they do it, comes down not to zoological altruism but to the need to raise and nurture a happy family of kits.

As a rodent, albeit Europe's largest, chisel-toothed and fierce when required, beavers are vulnerable to predators. In Europe, the beaver's natural predators include brown bears and wolves, whilst red foxes, otters and badgers can all prey upon kits. Most of Eurasia's other rodents have developed, against such attacks, the defence of a burrow. Long and deep, with an entrance no larger than its inhabitant, a burrow is the simplest means by which small rodents can, most of the time, keep their families safe from the larger animals that hunt them when they are asleep or their young still defenceless. Most beavers, on recolonising an area, may begin by building their burrows under the root plates of old trees, wedged into the banks of tributary streams. Over time, as the water levels rise due to damming by the animals, beavers become more ambitious – and will, preferentially, build themselves a new creation: a lodge. The lodge arises from one fundamental need: beavers prefer their front door, wherever possible, to be safely underwater.

A beaver's lodge, hewn from branches and deep-set cobbles, is so strong a fortress that even a bear would struggle to break in. Whilst its two or more front doors, must, for the beaver's safety, remain below water, at times, in drier summers, the front door can become exposed. During such a time, like humans repairing their guttering only when it begins to leak, beavers, on realising their front door is exposed, will get to work adding extra mud to the lodge, and the dams around, to retain more water. Beavers thereby raise the water level, until it covers the main entrance once more – and constantly adjust the water levels outside their home in this way.

Even when rivers or streams freeze over, beavers can escape from their submerged front doors into the water, swimming below the ice for up to six minutes before coming up for air. In the ceiling, a single air-vent allows hot air to escape and ventilates the lodge. Indeed, in the depths of winter, a gentle plume of steam in the freezing air, rising from a snow-capped beaver fortress, is a sure sign that its inhabitants are safely tucked up inside. The lodge allows beavers to eke out a living at the harshest times of year. It is their retreat from predators, the place where they will winter, and, in summer, where they will raise their kits.

From the safety of the lodge, a newly established family of beavers will get to work shaping the landscape all around. Dams, the most famous of beaver creations, often follow next. As any engineer will know, the first stages of construction are often the most fraught. Often, in order to get the foundations of a larger dam cemented into place, beavers will already have lessened the flow of the river by building smaller dams upstream, continually making new ones whenever water begins to come around the side of the one they built before.

A beaver's teeth are orange for the same reason that our blood runs red: their primary weapons are fortified with iron. This renders a beaver's teeth durable against mechanical stress, and also resistant to the acidic content of tree sap. Now, the forestry operation begins in earnest. Unlike in the cartoons, beavers rarely fell whole trees to build dams – although they most certainly can. Most often, the limbs and branches of smaller-diameter riverine trees, especially aspen, birch and willow, are hewn. Up to 80 per cent of trees stripped of their branches will survive, sprouting into dense coppices of new shoots.

Bearing in their mouths neatly clipped branches that can weigh as much as themselves, the beavers commence construction by diving to the bottom of the water. Here, they place, push and position branches and logs into the riverbed. The foundations complete, the rodents then start work on the main edifice.

Tree bark, leaves, rocks and plant matter are all brought to the growing dam – then zealously cemented with mud. Whilst beavers are not believed to use spirit levels during their work, most dams end up extremely level, from one side of the stream to another, often following the 'horseshoe' formation, beloved, in recent centuries, of human engineers. One of the reasons for this is that beavers can hear the sound of water escaping through an unlevelled dam, and rush to repair the damage. By the time the construction is complete, a dam in a typical river may stand, from the river's bed, anywhere up to two metres in height. Others are less dramatic affairs, shaping rivers as a collective, but each trapping only a small volume of water by itself, and thus creating a complex network of ponds.

The first purpose of a dam is to create a safe foraging habitat for the beavers themselves. In fact, a beaver pond, formed in the wake of a dam, is not unlike a castle moat – keeping the enemy out, and beaver society safe within.

The water will often stretch back from a dam for up to hundreds of metres, expanding to form a network of impounded ponds.

By creating such impoundments, beavers create alluvial highways through which they glide and forage. Swimming in water where there was once land, the rodents then drive home the advantage – digging deeper channels within the flood to allow them to penetrate further into the woodland – and access more food. Powerful in water yet shuffly on land, a beaver is safest when closest to the water's edge. Here, it can rapidly dive back into its element should a fox or, in increasingly large parts of Europe, a grey wolf, arrive on the scene. These riverine highways also allow beavers to use the surface tension of the water itself to transport far larger branches than they could ever carry overland. Each of these is then taken back towards the lodge. This, in turn, leads to the second crucial function of the dam. Dams are not just structures designed to slow the water's flow. They also act as the beaver's *larder*.

After a long spring and summer feasting on vascular plants, come autumn, beavers become busier yet, harvesting succulent branches of their favoured aspen, willow, birch and other trees in the same way that a huntsman or farmer stockpiles logs for the winter fire. Each of these is taken below water, and wedged securely below the burrow or lodge. Here, preserved in the cool still water, these branches escape the icy onslaught of the impending winter, preserved, like salted meats, in the beaver's winter kitchen. This submerged larder of branches acts as a crucial source of nutrition during the hardest of months. Each watery limb will be wrestled from the base of the dam and taken to the lodge, to be stripped of its bark there in safety.

Within a Eurasian beaver's cosy winter home, love is often in the air. Beavers will mate mostly between December and April, in preparation for the coming summer. By spring, having used this larder to defy the winter's course, they get back to work once again, renewing and repairing their dams, or moving to a new lodge if an existing one has become unusable.

Beaver kits are born from early May onwards, when the willows and birches are at their most feathered and lush, their sap richest and fullest, and aquatic plants most readily available, soft and ready for the kits to consume. Young beavers will remain with their parents for close to 20 months – kept warm and safe within the lodge for their first few months of life – before setting out to engineer their own place in the world. And over time, from the basic instincts of a rodent – to find food, stay alive and raise a family – emanates one of the richest habitats on Earth.

What beavers actually do is simple enough, but the outcomes of those simple actions are both extraordinary and far-reaching. So rich and complex is the world created by beavers that scientists are still unravelling it. An ever-growing number of species are found, year after year, to prosper from and even rely upon the presence of beavers. From the first dam and the resulting first impoundment of a river, nature's makeover begins. And it is the intrinsic make-up of the beaver pond that initiates this transformation.

Beaver ponds are warmer than many surrounding environments within a river's catchment – or, indeed, ponds that are not made or maintained by beavers. There are two key, complementary reasons for this. Firstly, by the beavers' dams slowing the water's flow, the pond's calm waters are warmer than those either up or downstream. Secondly, beaver ponds are warmer at the surface, because the bushes and trees around are continually coppiced or felled. As this opens up the canopy, and sunlight now glances directly down into calmed waters, accelerated photosynthesis fuels a frenetic growth of microscopic life. The surface waters of a beaver's pond, therefore, become an ecosystem, nature's equivalent of a Jacuzzi.

Plant and animal plankton abound in higher numbers here than elsewhere on the river, and their presence builds the invertebrate trophic cascade upwards from the bottom. The accumulation of organic matter within beaver ponds, coupled with the decrease in the rate with which the water flows, begins to create more sympathetic breeding conditions for stoneflies (whose nymphs feed larval dragonflies), caddisflies (prey for a range of species, including dippers) and mayflies. Generally, areas immediately upstream of a beaver dam will find higher concentrations of all these species than on un-dammed rivers. The stilled waters provide the fragile, early life-cycle stages of these species with the conditions needed to reach adulthood. On emergence, mayflies will dance above the water's surface to live out their short lives, whilst caddis will move downstream. Chironomidae (non-biting midges that transform detritus by eating it, and are eaten in turn by a host of small wetland predators) also live best in the warm sanctum of beaver ponds. In the USA, it has been shown that small invertebrate predators, such as wolf spiders and beetles, flourish in the margins of beaver ponds, attaining far higher density than elsewhere, owing to the abundance of warm-water midges. Warm water also increases the speed with which these tiny creatures grow. The surface waters of a beaver's pond, within less than a year of its creation, thus become places of super-abundance for micro-invertebrates.

As a result, a range of birds, including the tree-nesting goldeneye and ground-nesting teal, often find food around beaver ponds. As the adults bring their ducklings to the ponds, the abundance of small aquatic life fuels the growth of the young birds. Wading birds like green sandpipers, which breed in the disused nests of other species, often high

up in pine trees, commute through woodlands to find
shallow pool systems in which to feed and raise their chicks.
Whilst common in the beaver-hewn woodlands of Finland,
Poland or Estonia, these wading birds struggle in Britain to
find such habitats, although with the beaver's gradual return
across the UK, this may begin to change, especially in
northern Scotland where green sandpipers do still breed,
albeit in tiny numbers. Any wading bird on passage elsewhere,
dropping onto the side of a beaver pond, will find – in the
verge between water and wet wood – a veritable feast of
small insects.

Other larger, more dominant species, such as the mute
swan, or cranes in Estonia, can find beaver dams the perfect
place to nest. Their home is almost guaranteed to remain
above water throughout the nesting months, and their
doorstep swims and wriggles with the richest menu for
miles around. Other birds can rapidly appear and nest
beside beaver ponds as soon as a community of worms,
crustaceans, spiders and gastropods has developed. In
Tayside, the many fallen boughs found within beaver-
shaped loch-sides become fertile hunting grounds for
common sandpipers and pied wagtails. In Cornwall, water
rails – a species we most often associated with larger British
reed-beds, have appeared beside beaver ponds within years
of their creation, in valley bottoms that may not have
echoed to their squealing pig calls for centuries. All have
come to pay homage to a superabundance of miniscule
invertebrates.

Common frogs and both common and natterjack toads –
all of which have lost most of their British abundance in
the past 50 years – are also well looked after by beavers.
The warm, rich waters of beaver ponds accelerate the safe
growth of tadpoles, whilst the complex, sunlit structures
of dams afford excellent habitat for adult frogs and toads.
One study of comparable amphibians in Canada showed

that five times more wood frogs, and more than 20 times more western toads, were to be found in beaver ponds than in surrounding streams. Tadpoles, in particular, forage in the rich decaying plant matter left in the wake of dams. But most important of all is the life-giving presence of sunlight itself.

With the thinning of much of the vegetation surrounding the beaver pond, perfect feeding and breeding conditions for amphibians are created, too. Beaver ponds not only contain more amphibian young, but also far higher numbers of successfully metamorphosed adults. Great crested newts, for example, prefer warm, sunlit ponds, but also burrow into mud sediment to survive the winter; a substrate found in abundance at the muddy bottom of a beaver's pond. Other species prefer the role of stowaway. Slow worms, grass snakes and adders can all slither into the cosy upper crannies of a beaver dam. In one study, slow worms and grass snakes colonised rivers *only* in the wake of beavers.

Whilst the top layer of a beaver pond is warmer, benefiting a range of fast-growing aquatic life-forms, the pond's bottom strata is the opposite. This is an important duality within the structure of a beaver pond: its warm surface waters conceal another world, hidden from view: one that is cool, dark and still. Furthermore, the bottom strata of a beaver pond is as complex as the surface of a coral-reef; filled with nuanced monoliths of dead timber, rich in hiding holes. It is here, where many logs are stored in the winter larder, that a new ecosystem develops. Here, the submerged base of the dam, and the pond's complex log lining, is colonised by dragonfly larvae. By May, emerged adult dragonflies dance above beaver ponds, often in higher abundance than either up or downstream, their voracious, carnivorous larvae well fed on the pond's stock of water beetles, tadpoles and worms. But a beaver pond's subterranean depths hold greater riches than dragonflies. They are nature's oldest, natural salmon farms.

Whilst well known amongst American or Canadian naturalists as a wild guardian of salmon species, the role of beavers in protecting salmon stocks is much misunderstood here in Britain. But the means by which this is achieved is relatively straightforward – and owes much, again, to the beaver pond's design.

In all, there are no fewer than *five* ways in which a beaver pond's lower strata comes to act as a perfect crèche for colder-water species such as salmon. First, deep, cool waters allow young fish to hide from surface-hunting predators, such as herons. Second, the complex, log-strewn base of a beaver pond affords thousands of tiny hiding places for developing fish to grow. Third, due to the stillness of the water here, compared to the surrounding river catchment, the lower reaches of a beaver pond allow fish to develop faster. Now, instead of wasting critical energy fighting the flow of the water, young salmon can spend more time fattening up on the food trapped by a beaver dam in the first place. Fourth, the abundance of trapped invertebrate food here is far higher than either up or downstream – and thus the young fish develop more rapidly. Fifth, and finally, as if that were not enough, successive beaver dams, slowly reforming straight rivers into a helix of bends and eddies, create more shady corners wherein younger fish can hide. Indeed, by the end of their time within a beaver's catchment, young salmon are now in prime condition to continue their travels downstream to the sea – having been given the perfect start in life!

Once fish have developed within a beaver's pond or wider river catchment, they can, and do, escape. Often, this is simply by swimming through the water that makes its way around the edge of each dam, forming a new mini-tributary. Equally, migratory fish, especially on their return migration, are adept at leaping beaver dams entirely. Indeed, it has been speculated that beaver dams, once being far more common than waterfalls or weirs, may have taught salmon to leap in the first place!

Whilst some angling lobbies in Britain continue to argue that beavers threaten salmon, most of the world's ecologists, and many of its anglers, remain puzzled by this unusual perspective, which shows, amongst other things, a total lack of understanding of fish – and how they evolved. Beavers and freshwater fish shared the same landscapes for millions of years, and beaver ponds, and dams, act as some of nature's most important fish crèches.

As early as the twelfth century, when travelling ecclesiast Giraldus Cambrensis walked the Tywi in Wales, in 1188, he noted that, 'The church, mill, bridge and salmon leap, an orchard with a delightful garden, all stand together on a small plot of land. The Teivi being the only river in Wales which has beavers.'

Centuries later, on the other side of the Atlantic, Canada's Hudson's Bay Company would learn the hard way what a swift reprisal against beavers meant for salmon stocks. In 1818, following a government dispute over trapping rights, the company ordered its trappers to wipe out all fur-bearing animals in the region – and the beaver was one of the first. The following year, in an era long before commercial fishing or pollution, salmon stocks plummeted. Indeed, in most of North America, the role of beavers in salmon preservation is well known. In the Western Cascades of Oregon, scientists working with the Tulapip Tribes have sought to restore beavers as a matter of urgency, in their bid to save the coho salmon from regional extinction. In 2004, the University of Washington accepted that without the help of beavers, salmon restoration would have extremely 'limited success'.

One of the more surreal objections to the restoration of beavers here in Britain, has been that salmon, or other migratory river fish, cannot jump over their dams. In terms of simple evolution, however, the argument does not hold up. Until a few thousand years ago, there is little evidence that beavers were hunted as a major food source – and thus across Eurasia and North America, every river, stream and wetland would have been governed and reformed by beavers. Had salmon been unable to pass upstream, or downstream – either by clearing dams or through diversion of their routes – they, and many of the world's fish, would have become stuck, rendered extinct and thus removed from the natural order, millennia ago! Funnily enough, this didn't happen.

The reality is that beaver dams are not, unlike a concrete dam of human creation, impermeable. They slow, divert and meander rivers – but do not stop them in their tracks. Close observation of any dam system will reveal natural 'fish passes'; smaller channels that work their way around the dam's edge. These passes are not, of course, created by well-meaning beavers, but are in fact new tributaries, diverted by the beaver's dam. Water always finds a way – and a beaver dam inevitably allows for the safe passage of water, and fish, downstream. By the time far larger salmon, as adults, return upstream, they have the power to negotiate these passes with greater ease, as well as the fitness to jump over many dams – a behaviour commonly observed where beavers and salmon coexist more widely in the USA and Canada.

Finally, in studies carried out in Lithuania, beaver-dammed rivers not only saw an increase from nine to fifteen species of freshwater fish, compared to rivers with no beavers, but also enjoyed richer fish *abundance*. In particular, chub, trout and perch were found to flourish in beaver ponds. And this profusion of fish benefits both human and avian anglers. Kingfishers, herons, cranes and, in mainland

Europe, ospreys, have all been observed to head to beaver ponds in the reliable hope of a take-away fish supper. And so, whilst it can be hard to credit a rodent with successful fish-farming, the widespread restoration of beavers to British rivers will, in fact, greatly increase its abundance – and diversity – of fish.

The effect of beavers upon their fellow mammals is perhaps even more surprising. The abundance of fish and amphibians in the wetlands that trail back from beaver dams gift unique hunting advantages to species as large as otters. Furthermore, old or abandoned lodges make the perfect summer holt, as do those lodges of beavers built into the sand shores of river-banks. In the Mazowsze region of Poland, otter densities run twice as high in territories occupied by beavers. Even the holes that beavers cut in the ice allow otters to surface and breathe as they conduct winter raids below snowy rivers.

Bats are one of the more surprising orders that qualify for beaver benefits. By creating ponds and, eventually, wet meadows in woodland, free from canopy cover overhead, beavers unwittingly gift to bats the perfect hunting grounds. At Cornish beaver ponds, up to nine of the regular 13 British bat species have now been regularly observed hunting. Without clusters of branches overhead, and with the rich, insect-giving waters below, bats can hunt freely without interference to their echolocation. In northern Poland, it was found that bat species familiar to our own shores – pipistrelles and noctules – darted far more often over beaver ponds than adjacent areas of water. The beaver ponds made such flight paths clearer, and their warmth fed the night with a rich abundance of airborne prey.

More surprising again are some of the beaver's larger, terrestrial shadows. German wild boar, it was found, were

drawn to most of the studied beaver ponds in one river
system; scavenging water lilies from their drier fringes.
Badgers and martens have readily moved into recently
abandoned lodges. Throughout the ponds and banks of
beaver impoundments, one of our most beloved and
threatened riverside characters, the water vole, finds its
perfect summer home.

In fact, it has been found that a host of familiar smaller
mammals will often colonise areas *only* in the wake of beaver
activity. Weasels, which enjoy the complex stone walls and
rabbits of upland farmland, are equally at home hunting the
bouldery shores of beaver dams for prey. Bank voles, an
important food for tawny owls, are perfectly adapted to
become unobtrusive neighbours within a beaver lodge. But
there is one animal, another ghost of our shores, that may
yet benefit from beavers in the ever-growing wetlands of
Britain in the centuries to come, were it, too, to be
reintroduced – and that is the European elk. Studies in
Finland have shown that elk benefit from the coppicing of
beavers, which greatly enhances the overall biomass of
willow and aspen forage available to them. Yet even amongst
our more familiar British mammalian fauna, all the cast of
The Wind in the Willows, from Badger to Ratty, are thus
provisioned for by that other endearing character and the
engineer of the willow itself: Beaver!

Whilst beaver ponds and dammed rivers have, in their own
right, the power to sustain and save hundreds of species,
the dams and resulting impounded ponds are merely the
beginning. As time and beavers conspire, different
landscapes, outcomes and lives emerge – and new players
enter the game.

By creating dams and shallow water in woodlands,
beavers increase the abundance of the very tree species that

they need to eat. In an endless cycle, willow, aspen and
birch, in particular, colonise the damp soil. Many are
coppiced by beavers but most survive, growing to a ripe
old age but in bushier form. Then, in death, the trees create
new habitats as they decay and eventually fall. Around and
around, the cycle of growth in our riverside woodlands is
never-ending.

Many conservationists and traditional foresters, from
those creating willowy bankside habitats for warblers to
those managing woodlands for nightingales, lament the
decline in recent years of coppicing in the British
countryside. Coppicing, put simply, enriches biodiversity by
turning straight trees – of surprisingly limited use to most
wildlife – into denser, bushier and wider ones that often
proliferate with a greater number of nesting sites – and a
greater abundance of small invertebrates to boot.

By repeatedly cutting back resilient trees like willow or
hazel close to their roots, creating a 'stool', woodland
managers promote the growth of dozens of new shoots.
These rich, billowing networks of stems afford more cover,
and more food, for many nesting birds. Human coppicing is
in decline; costly and time-intensive. With beavers in the
landscape, it becomes common once again, and is achieved
free of charge. By repeatedly gnawing off medium and large
branches, more often than tackling the girth of entire trunks
(too large to carry), beavers effect the growth of thousands
of new shoots. These new scrublands rapidly develop into
habitats of great ecological value.

Many of Britain's most rapidly declining songbirds nest
and sing in bushy, complex worlds. For these species to
persist in abundance and diversity, and spread across our
currently impoverished landscape, it is clear that we will
need beavers established once more across the British
landscape – and not just in tiny pens, where their wild
glory is confined within zoo-like enclosures. Over time, if
wild beavers were to colonise large areas of the British

countryside, the list of avian beaver beneficiaries would
grow year on year.

The reed bunting sings from vigorous riverside bushes
beside reeds or wet grassland. The strange, mouse-like
grasshopper warbler is fussier yet, tied not to a habitat but to
a *moment*, when young saplings burst upwards through dense
scrub and tall grassland. In many areas of Europe, bluethroats,
which have yet to colonise Britain, breed best where rich
scrub develops within marshland. In Poland's Biebrza
Marshes, the dense coppice of beavers forms the natural
habitat of birds like corn buntings and tree sparrows. We
have forgotten that millennia before humans invented the
hedgerow, beavers had got there first.

Another charismatic singer lost from the British landscape,
the marsh warbler (which can impersonate a staggering
range of other birds, from starlings to golden orioles), sings
in a complex world of wet herb meadows and dense bushes.
In Britain, the nineteenth-century practice of planting osier
beds (willows repeatedly coppiced, to produce withies for
baskets and fish traps) saw an upsurge in marsh warblers.
Now, that practice has all but disappeared. Beavers, however,
create osier beds, and herb meadows, on a daily basis, and on
the continent, these habitats are often used by breeding
marsh warblers.

In the beaver-coppiced riversides of eastern Poland,
you will commonly find the willow tit, a bird on course
for rapid extinction here in the UK. Willow tits achieve
their highest densities in a rotting chaos of damp, maturing
scrub. They excavate nests in rotten stumps – a rare
commodity in many of our sanitised woodlands, yet a
resource that the beaver provides in profusion. It has been
observed by naturalists that willow tits often nest very
low; just a metre or so above the ground. This allows them
to move into the birch or willow stumps left in the wake
of a chomping beaver. Willow tits flit through the dense,
fine branches of birches, thorn trees and willows in search

of small moth caterpillars; the beaver's coppicing creates such densely packed branches because, far from killing willow trees, beavers turn a large amount of these into bushes, as willow spindles out into its richest, fullest form. Finally, willow tits spend more time than any other bird feeding, and often nesting, in elder. Elder is toxic to most animals – and beavers leave it well alone. So willow tits in a beaver territory find the perfect combination of feeding and nesting sites, and, in all probability, evolved to exploit and excavate such habitats over many thousands of years, being co-evolved, like all of Britain's wetland species, beside beavers.

In many areas of eastern Europe, nightingales not only remain common but also attain high densities on the edges of beaver-coppiced wetlands. But in Britain, many ecologists have observed that nightingales face a distinct problem: their habitat only lasts for a few years. The succession of young woodland soon turns their dense scrub blanket into dense, 'leggy' trees – which no longer create the shady world in which nightingales sing, forage and nest. As a result, in Britain, much time and money is spent upon 'arresting' succession, through repeatedly cutting scrub back. One of the most powerful actions of the beaver, however, is to ensure that the development of a landscape is not linear but *cyclical*.

By repeatedly felling young trees, again and again, beavers effectively reset the clock on entire habitats. In a beaver wetland, there are always newly felled trees, older trees, and newly coppiced ones, existing at the same time, side by side. Nightingales, therefore, need never move out – there is always enough new scrub to go around. The ability of beavers to effectively halt ecological time, therefore – defying a closed canopy, maintaining open ponds and resetting the scrub landscape every year – is unique within the animal kingdom. Not only do beavers create habitats for scrubland birds, but they also allow for permanent suitability for species such as nightingales; something human conservationists have always struggled to achieve.

At the present time, so nascent, penned and, in some cases, persecuted, are Britain's foundling beaver populations, that their widespread effect on birds has yet to be enjoyed – but this is no idle speculation; the scientific proofs from other countries, where beavers persist at a landscape level, is overwhelming. In Wyoming, studies of the diversity of low, bushy vegetation that develops in the wake of a beaver dam has revealed an explosion of abundance in declining songbirds. Now, many American and Canadian ornithologists are hastening to put beavers back.

Just as beavers enrich a woodland's early years by creating dead stumps, dense bushlands and a rich diversity of woodland, so, decades later, these actions reverberate in the now *mature* forest. Whilst many aspen are felled by beavers, those that are not face less competition from their fellow trees – and prosper under sunlight. Any forester will explain that thinning trees greatly helps the development of a woodland. This is because woodlands have evolved to be thinned – with beavers, around waterways, originally doing much of that thinning. By taking out large numbers of younger trees, beavers accelerate the passage of the survivors to gianthood.

Beavers never linger too long in one, precise area. In abandoning an area, for example, beavers will leave large aspens to grow into giants. Decades later, they may return, flooding the same woodland. Now an aged tree, already at the end of its living years, the aspen's rot will be accelerated by the new flood gnawing at its roots. By the time beavers have shaped landscapes for a century, their growing effect on the natural world is that of time on a cask of ripe French wine. Old willows, poplars, birches and alders, standing in swampland, attain yawning rotting forms. Woodpeckers like the white-backed, which in historical times may once have bred across Europe, are tied almost entirely to such extensive rotting lands. The lesser spotted woodpecker, now vanishing across Britain, inhabits a damp, rotten world of rotten trees reaching the end of their lives. In the USA, several studies have found that by accelerating the decay of old trees through flooding, the action of beavers creates a higher density of nesting woodpeckers than woodlands where beavers are no longer present.

In turn, these old woodpecker holes, some enlarged by other animals, others by the elements, become some of the most desirable apartments in the canopy. Tits, flycatchers, robins, smaller owls and tree-nesting ducks can all use the holes in various stages of disrepair, as can hornets, bees, wasps and spiders. Then, at last, succumbing to age, water and gales – these swampland apartments fall, as the wind blows down the willows.

Now, deadwood supports infinitely more life than the living. Into the crumbling creviced forms of fallen swampland trees moves an army of beetles and fungi. Longhorn beetles feast on fallen aspen. But close by, new beaver ponds, which increase beetle diversity by up to a third by providing both water and rot, now work in tandem with the process of decay. From the beaver pond, some beetle species will expand their range into the ever-growing myriad of decaying habitats around. Now, beetles raised in one part of the beaver landscape begin to attain vital roles in breaking down timber within another. Soon, surprising new visitors come to pay homage to the fallen forest. In Poland, brown hares and mountain hares, in autumn, winter and early spring, lollop in to feed on fallen boughs. Fungi creep in a beaver's wake, boring and gnawing through tree stumps.

As beavers abandon their dams over time, and what were once ponds give way to meadows, a new dynamic is born.

Wet meadowland systems grow to prosper. As a beaver pond drains, and water sinks into the soil, a new generation of grassland is born. In eastern Poland, anthill-raiding wrynecks – which prefer wood pastures with short grassland – move into these habitats. And this process of habitat creation becomes cyclical, self-sustaining and continually evolving. In many cases, the small-scale variety (or heterogeneity) of such landscapes come to exceed anything that human conservationists have been able to achieve.

Wet meadows, for example, are a habitat that, for a long time, conservationists in western Europe have sought to recreate – but with very limited success. Such endlessly evolving places – being more *moments* than habitats, and requiring many different levels of water, in one place, across the season's course – have proven difficult for human interventions to mimic. As a result, a host of species that would once have haunted our beaver meadows have, over centuries, vanished from much of the European landscape.

Of Britain's lost butterflies, the large copper was once the glowing beacon of our wetlands. A deep ember burning bright in lush, damp grass, the large copper breeds best where water dock grows in damp, low-lying verdure. By 1852, its populations reduced over centuries of drainage, it vanished from Britain – the earliest recorded of our butterfly extinctions. In 1927, conservationists reintroduced the large copper to Woodwalton Fen, in Cambridgeshire, where, with careful management, it survived only until 1969. On reading conservationists' despairing descriptions of the butterfly's preferred habits, it becomes evident that the large copper has evolved to be a fussy customer. The coppers lay their eggs on dock, in low-lying areas that must not become flooded in late spring, else the caterpillars all get washed away. The larvae survive only in areas that are damp, but not wet – thus preferring the margins of fens. Males prefer to feed in warm, damp depressions in the grass, but couples court in rich tall stands of herbs. What's more, large coppers live only in highly diverse grasslands, where nectar abounds, and thrive only across large landscapes.

As conservationists have found, creating such a mosaic is a battle rarely won. Yet in Poland and Estonia, where the large copper, and beaver, are both increasing, a beaver meadow achieves all of the above. Dams, even old ones, holding and slowly releasing water, create the perfect damp conditions where copper larvae will not be washed away. Old beaver ponds, slowly overgrowing with herbs, provide courting grounds. Rich alluvial soils fuel the growth of daisies, on which coppers spend much of their time feeding. The sunlit depressions of old ponds afford warm lush glades. And in the ever-changing world of the beaver, all of these micro-worlds exist side by side. Like so many animal orders whose lives we have tried to prescribe, the large copper, having evolved beside beavers, is perhaps, naturally, a beaver butterfly. It is not the only one. In the boglands of Wisconsin, lepidopterists have found the Gilett's checkerspot is tied to 'cyclical habitats' but breeds best in those areas where beavers are in charge.

In Britain, in spite of a long-standing passion for butterflies and their preservation, we have one of the worst track records in Europe of actually reversing their declines. Most of our native butterflies require a myriad of tiny habitats knitted together – with herbs, grasses, flowers and many types of scrub and woodland all mingled within the same landscape. The high brown fritillary whizzes over woodland clearings, feeding in sunlit glades but laying its eggs on dog violets, which benefit from moist soils. Wood whites flit like little ghosts through sunlit herb-lands at the woodland edge. Across Europe, the chequered skipper flickers on the verges between sunlit wet grassland and shaded woodland, requiring water trapped in the soil to grow its favoured food-plants: moisture-loving grasses. We can perhaps trace the beaver's tooth-marks through the habits of such imperilled butterflies – many now refugees in a denuded world; a world we garden, at great expense, to maintain the last of each species here in Britain.

Whether specialised or common, all butterflies require a diversity of plants to cater for their remarkable life cycles. In a 12-year study conducted by the University of Stirling on the beavers of Tayside, in Scotland, it was found that the number of plant species increased by 50 per cent compared to nearby areas from which beavers were excluded. Rebuilding trophic cascades from the bottom up, the action of beavers is a far better guarantor of butterfly diversity than anything we can replicate.

In Britain, of the many habitats we have failed to save, or reinstate, the ephemeral world of lightly flooded grassland has eluded conservationists the most. The black tern, lost from the Fens by the 1820s, once nested communally on shallow floodplains that neither flooded, nor drained, too fast. The black-tailed godwit and the ruff, feeding in shallow water and displaying in the open, hide nests in damp grasses that grow with the season's course. The glossy shoveler, with its iconic spatula of a bill, sifts the surface of shallow waters then flies to its nest hidden in long grass. The redshank, now scarce in much of lowland Britain, breeds in the dampest of grasslands that are not, quite, flooded. The spotted crake, which remains common in Europe's untamed river valleys, breeds only when shallow water lies over sedge beds, but requires dry islets of sedge within which to nest. The jack snipe winters in similar habitats in Britain, and has been flushed regularly from beaver-ponds in the south-west. Corncrakes, whose lush damp grasslands risk being overgrown by scrub, are perpetually given new homes as beaver ponds dry out. Termed 'farmland' birds in western Europe, they remain common, in eastern Europe, in damp woodland clearings. In the beaver-crafted woodlands of Poland, Estonia and Latvia, honey-buzzards and black storks are regular visitors to wet forest meadows, where they feast on amphibians. Given a chance, all of these are beaver birds.

Our uplands, too, are bereft of many shallow pools that beavers would create. Species like cranes once haunted our moorlands, but even now, regaining lost range in the UK, they would struggle to nest on many of our fast-draining moors. Wading birds like curlews would benefit from the shallow, upland wetlands that beavers create and the profusion of small invertebrates found in beaver ponds which, in drying, would turn into curlew-friendly upland meadows.

Other British birds have become, in recent times, highly dependent on managed human landscapes or even man-made islands – but would not have evolved within such habitats. The avocet, which here at home often nests on scrapes with human-dug pools, is, in the valleys of eastern Poland, adapted to nest on shingle islets in the centre of large rivers. Many of these, in rivers like the Bug, are sediments washed, over time, against beaver dams, until shingle islands are created. But only large-scale beaver action, over decades, if not centuries, is capable of creating such large-scale results. The greater hold beavers have within the landscape, the more structures, habitats – and species – they create and provide for.

If this was all beavers did, that would justify their widespread return as a matter of ecological urgency. But that is not all that beavers do. Given the chance, given the space, beavers protect the lives of another species – *us*.

Had you happened to be standing beside the remote Frank Church River of Idaho, USA, sometime in 1949, you may have been accosted by a rather unusual sight. Looking up into the sky, you may have seen a small squadron of beavers parachuting gracefully through the air. Checking yourself for inebriation or heatstroke, you may, after a while, have seen the beavers land in crates on the ground, with the male, and female, a short distance apart. Escaping successfully from their beaver boxes, you may then have watched the confused rodents re-orientate themselves and head down to the river.

This unusual but effective conservation experiment to relocate beavers was not the first of its kind. For almost the entire decade before, the state of Idaho had come to realise that beavers could not only enhance ecosystems, but also carry out two functions that humans simply could not. Firstly, they could reinvigorate Idaho's parched farmlands, rendering them productive once more. Secondly, beavers could prevent homes from flooding.

As long ago as 1939, the Interior Department had come to realise that whilst in a few specific areas, too many beavers were causing a backlog of water on the land, in others, eroded, silted-up catchments were no longer holding water at all. The department sought to redistribute the animals, writing:

> The value of the North American beaver … lies as much in his teeth and his temperament as in his fur … By the end of last season, some 500 beavers were busily damming streams under Government supervision … With hundreds of arid Idaho acres already reclaimed by silt-catching beaver dams, Department of Interior experts look forward to using more beavers in Oregon and California. Cost of trapping and transplanting a beaver: $8. Estimated value of one beaver's work: $300.

Over the next decade, beavers were redistributed and relocated across the state of Idaho – and as they were, the entire landscape changed. Large areas of farmland, where the soil had become simply too dry to plant crops, became reusable, as the beaver's actions helped to re-saturate the land and return its nutrients. In 1941, a feature in *Time* magazine detailed the gratitude of residents to the beavers of Salmon, a small Idaho town, for 'saving the city the cost of a dam', as the beavers' actions protected it from flooding. For the best part of 80 years, it has been accepted by the majority of people in America and Canada – whose national symbol is the beaver – that the employment of animals to provide for and protect human beings is, ecologically and economically, essential.

In recent years, the reasons for beaver reintroductions in the USA have proliferated. Facing the effects of climate change and the growing risk of wildfires in the hills of the Pacific Northwest, scientists have realised that decreased snow-retention, and increased snow-melt, will rapidly drain rivers of their water – and salmon – come the summer. By returning beavers to the upland rivers, scientists are hoping to delay and mitigate some of the worst effects of localised climate change, as beaver dams re-saturate the streams by damming; slowly releasing the water over months, rather than days.

The actions of beavers in preventing flooding are as much a human preservation service as an ecosystem one, yet to many in Britain, these actions remain much misunderstood. And, in fairness, the issue is not entirely straightforward.

In short, upstream of a dam, beavers can cause *some* flooding as the wetland backlog behind their dam fans out slowly across the area around a river or stream. Such flooding is slow and gentle in nature; it can threaten certain crops, like potatoes, growing near the water's edge, and beaver dams built adjacent to homes or gardens will rarely be welcomed. However, those working with beavers in Britain have made an important discovery. Allowing just *20m* in total, around a river's girth, allows beavers to work their wonders with zero impact on the farmland around. So much for the flooding *upstream*.

Beavers, it has been proven, many times, prevent *massive* and potentially life-threatening flooding *downstream*, because of the complex ways in which their dams hold and release water. A study by the University of Exeter on a small, fenced population of beavers, recently reintroduced to Dartmoor and confined to just two *hectares*, showed that 13 small beaver ponds were able to retain and hold back more than 600m^3 of water. Rather than pelting off the hills within a day, the water was gradually released over the course of months. Whilst such revelations would be considered a distinct non-event in several countries, from Canada to Sweden, it still comes as a surprise to many here in Britain that, in spite of building dams, beavers do more to prevent harmful floods from hitting our homes than they do to cause the same.

With any degree of practicality, beavers might also be looked upon as powerful assets to the majority of farms, especially upland farms or mixed farms, for the simple fact that they keep water on the land, providing drought-proofed aquifers. There is nothing to say that each beaver pond need be a pristine entity. Indeed, any enterprising farmer would not only actively encourage beavers on their land, but also give them their 20m 'buffer' needed to dam in peace, creating – from an unusable, narrow torrent – a veritable miniature reservoir, some of which could be drained to irrigate crops, feed animals, or both. This is why the concept of paying landowners to 'farm' beavers is also an entirely sensible idea.

In most areas of Europe, the philosophy that most beavers, most of the time, are here to help, has pervaded the actions of government and landowners alike. Populations driven to the brink have dramatically recovered and reintroductions have become commonplace. Across northern Europe, the life-giving effects of wild beaver populations have been restored to the land – everywhere except in Britain.

In spite of its unique ability to create life, save species and keep us safe, the beaver's struggle to return to Britain – as reintroductions remain painfully slow, and animals in Scotland continue to be culled – reminds us of how disconnected we have become from the life-giving forces that we need the most – and the respect for animals we must regain, if we are ever to repair our damaged land. Only as beavers shape our water, our ecology and our lives once again will the ancient rhythms of the countryside return, and a chorus we never knew was missing will be sung once again.

CHAPTER FOUR

Whales

More orderly, ornate and fine than any pendant or bracelet, the shapes of fossilised diatoms, crystalised under a microscope, appear too perfectly designed to be natural. Some are silver pendants; others gold purses, sapphire-studded plates or emerald clasps. Diatoms resemble ribbons, bows, stars or lightning bolts. They seem for all the world like the product of a long-forgotten metalworking genius, yet the symmetry and balance of a diatom's natural form is arguably even more intricate than the finest human jewellery.

In the Victorian era, scientists collected diatoms from marine deposits, with the main site being at Toomebridge in Northern Ireland. After boiling these deposits in hydrochloric and then sulphuric acid, 'diatomists' were left with a fine, pale sediment. Sifting this under a microscope, needle in hand, they isolated the diatom fossils from the saline brash. Then, taking days, sometimes weeks, the collectors would mount the diatoms on glass slides. Adhesive, then resin, held the diatoms in place, until at last their true beauty was revealed. Klaus Kemp, one of the last diatomists alive, reflects that it still amazes him how 'something so small can be so geometrically correct'. In their living form, diatoms are marine phytoplankton. They sail through the ocean, each sealed within a wall of silica – the microscopic equivalent of tiny boats in glass bottles. These silica 'frustules', the hard but porous walls of the cell, are what give diatoms their range of extraordinary, perfect shapes.

Though considerably less well known than the giant trees of the Amazon, diatoms are even more invaluable to the planet, contributing as much to photosynthesis as all our terrestrial rainforests combined. Tiny marine algae, individually no greater in size than half a millimetre across, diatoms lock away over 20 per cent of atmospheric and ocean-based carbon dioxide, and are responsible – in spite of recent human activity – for maintaining the relatively low levels of carbon dioxide in our atmosphere.

Not all phytoplankton are created equal, and diatoms possess a uniquely powerful biological engine – a CCM, or CO^2-concentrating mechanism. Highly permeable cell membranes maximise the area in which carbon dioxide can be absorbed from the ocean and into the diatom itself. The carbon migrates from the cytoplasm, the living material within the diatom cell, and into the chloroplast – an organelle within the cell, where photosynthesis takes place. Here, it is believed that within a structure known as the pyrenoid, carbon dioxide is broken down and reconfigured. As each diatom photosynthesises, oxygen is released. In total, at least 25 per cent of the atmospheric oxygen we inhale is exhaled by diatoms each year. Together with their fellow microscopic ocean plants, diatoms and all other phytoplankton in the ocean are thought to contribute over 70 per cent of the oxygen that makes it into our atmosphere – and our lungs. Quite simply, without ocean phytoplankton, land-based humans would not be alive.

When diatoms die, however, they sink through the ocean dark to the sea bed – effectively locking away the carbon dioxide that they have sequestered, and taking it out of the atmosphere entirely. Diatoms, some of the world's most abundant phytoplankton, therefore play a vital role in keeping our planet balanced – and alive.

If locking away carbon were not enough, they also power 40 per cent of the ocean's primary productivity. Floating gently at the bottom of the ocean food chain, diatoms are drifters; borne through the ocean on the current; nomads who harvest light energy from the sun, like plants, to photosynthesise. But they do not act alone. Another powerful single-celled algae, the dinoflagellate, is particularly active within our warmer temperate waters. Dinoflagellates too are phytoplankton. Many migrate vertically each day from the depths of the ocean to the surface, capturing carbon dioxide and then falling with it back into the deep. Both diatoms and dinoflagellates form the basis of the ocean's gardens. Where the phytoplankton are absent, the vast majority of ocean life is absent too. This is because phytoplankton feed, in turn, the smallest and most abundant grazing animals on our planet – zooplankton.

Zooplankton, in turn, are a varied assemblage of microscopic nomads. They include the tiny phytoplankton-eating copepods – microscopic aquatic crustaceans – but also the ocean's most abundant mini-predators. Zooplankton are conduits, transferring the energy of the phytoplankton, on which they feed, upwards through the trophic ladder, as they, in turn, are eaten by small fish. Of the many species that feast on zooplankton, one group in particular – the sand eels – are of paramount importance to the one group of birds for which the British Isles is famous worldwide: our seabirds. Yet the significance of sand eels can be hard to fathom. Until we see them, each summer, arrayed in the mouth of a puffin.

As a child, I always thought that puffins looked concerned. While the term 'cute' is more often used, the dark area above their eyes arguably gives them an expression of continuous apprehension. The black liner below suggests that they are short of sleep. Puffins look as if they have a lot on their mind – and, as a child, the reason for this seemed apparent to me. On returning to their nesting colonies, puffins arrive with not one but often a dozen or more sand eels perfectly lined up inside their bills, a feat both taxing and tiring. The maximum number of fish recorded in an Atlantic puffin's bill is 62. It is easy to imagine how a puffin catches the first. But what about the rest?

Unlike some seabirds, puffins do not generally forage far from their nesting colony. A tubby body and short whirring wings do not bestow upon puffins the wandering grace of the great albatrosses of the Southern Ocean, so they rely instead on the richness of the seas beside their homes. On many occasions, flocks will fish together in the same spot, often along tidal fronts. Here, puffins engage in a technique known as pursuit-diving: sighting their prey from the surface and then chasing it underwater. Filmed from below the waves, puffins first bob like rotund bath ducks, seemingly devoid of purpose or direction. Then, they snorkel, turning their heads below the water. Then, in a moment, these seemingly clumsy clowns transform into agile torpedoes. Suddenly, with the feet acting as a rudder, wings that make for comical flight become the turbines driving a deadly turn of speed. Trailing bubbles, puffins dart and turn in the water with a finesse that few would credit, seizing sand eels and other small fish with abandon. They have less than a minute to do so. Unlike penguins, puffins are not cut out for long-haul dives.

Once a puffin has caught its first fish, two adaptations allow it to hold onto the rest. The end of a puffin's tongue is rough and coarse. This sandpaper-like surface grips the fish, whilst the tongue deftly flicks it to the back of the puffin's wide gape. Here, backward-pointing spines fix the hapless catch in place like a plate in a dish-rack, allowing the bird to dive for a second, a third and a tenth catch without losing its hold on the first. The form of a puffin, perfected over millions of years, is based around extreme marine abundance. Whereas larger predators such as falcons or hawks may hunt extensively for a small number of calorific catches in a day, puffins hunt *intensively* – harvesting the waters close to home for a huge volume of sand eels, in successive fishing expeditions.

Not every sand eel brought back to feed a puffling – a plump downy mass shuffling around in its seaside burrow – will make it as far as that intended hungry mouth. Across the north and west coasts of Scotland, puffins must contend with one, sometimes two piratic species of bird, as finely adapted to rob the puffins as the puffins themselves are to hunt below the waves. Both great skuas, or bonxies, and arctic skuas are the muggers of the seaside. The arctic skua is an obligate kleptoparasite: it cannot help itself but steal. As much as cuckoos favour particular hosts in which to lay their imposter's egg, the survival of the arctic skua is tied largely to just four species of seabird, which it will harry remorselessly across the summer months: the puffin, arctic tern, kittiwake and guillemot. These are the sand eel carriers. Hanging in the air on falcon-like wings, arctic skuas wait for puffins and guillemots to come whirring in. Stooping suddenly, they harass the hapless fishers through the air towards the waves. Dropping or regurgitating their catch, the mugged seabirds must return empty-billed. After a short pause for breath on the cliff-top, Britain's haggard puffins, undeterred, must then head out once more, to land another catch of fish.

Skuas are the highwaymen of the sky. For millions of years, like puffins, their adaptation to feed upon sand eels has paid off. So assured has been the abundance of sand eels that not only whole seabird colonies but whole marine migrations have come to depend upon them. Arctic terns depart Antarctic waters in late March. In a staggeringly swift 40 days, crisp angled wings power the terns 26,000km north to reach coastal breeding sites in Britain. Here, alongside kittiwakes, puffins, guillemots and shags, they will join the sand eel hunting parade. Yet in recent decades, these migratory gambles have started to unravel.

Whilst our seabird colonies may seem multitudinous today, especially compared to the mere relics of bird populations left on much of the British land mass, these too are relics of their former selves. In fact, many seabird cities have slowly depopulated over the last century, as might our own towns, in the wake of a slow but continuous famine.

The Reverend Macaulay, visiting St Kilda in 1763 once likened its returning lines of puffins to hordes of locusts. As late as the 1890s, up to 90,000 puffins were harvested by hunting a year without wiping the birds out, but by the year 2000, the total population remaining on the islands was just 140,000 birds in total. In the last two decades, the rate of decline in our sand eel feeders has worsened dramatically. Even the wild, raucous reaches of places like St Kilda are not immune to the dwindling bounty of our seas. Shetland and Orkney, too, have lost almost 90 per cent of their sand eel-dependent kittiwakes since 2000, St Kilda even more. The arctic tern and the shag have been added to Britain's growing list of red-list species of conservation concern. The daring piracy of the arctic skua may be coming to an end; fewer than 600 pairs now patrol the cliffs of Scotland. As the sand eel feast diminishes, our seabird colonies fall silent.

The marine vacuum cleaners of zooplankton, sand eels act as the energy bars of British seas, transferring the protein of microscopic grazers up the trophic cascade to our seabirds, seals and larger fish. As befits their name, sand eels bury themselves for most of the day in the sand, emerging at dawn and dusk to feed. They require sandbars as a refuge, so our seas are not uniformly rich in these creatures. Sand barriers, such as the Dogger Bank, provide these fragile fish with critical hideaways. Most often, the choice of a stretch of coast by nesting seabirds, or a cloud of kittiwakes seen far out at sea, betrays where such sandbars lie. Being energy-dense, species like the lesser sand eel, the commonest in our waters, are still fished in large quantities each year, mainly from the North Sea: we turn them into animal feed and fertilisers. In turning sand eels, fished from places like the Dogger Bank, into the secondary components of factory farming, we drain from the ocean the lifeblood of our seabirds. But when it comes to sand eels, whose fisheries have declined in recent years, human predation is actually not the greatest threat.

Short of tiny marine grazing animals to eat, the hapless sand eel has, in recent years, been further depleted by fish we are more familiar with, most often in their frozen, lifeless forms, such as haddock and cod. It has been calculated that if haddock stocks alone were to recover, they would harvest as many sand eels from the Dogger Bank as the Danish fishery did at the height of its operation. What's more, mackerel, which have increased in recent years in the North Sea, not only prey on sand eels but also compete with them for food such as copepods. Such fish are the ocean's mesopredators – the smaller hunters of the ocean, the equivalent of foxes on land. In an undisturbed food chain, fish like mackerel and haddock are often hunted or outcompeted by larger beasts. And herein lies another problem. The saying goes, there is always a bigger fish. But in British waters, most often these days, there isn't.

In the early 1770s, the poet and encyclopaedist Oliver Goldsmith described hordes of seabirds, porpoises, sharks and dogfish – the 'millions of enemies' that chased and corralled Britain's super-abundant herring run as it came ashore. Goldsmith refers to harbour porpoises being so abundant that 'they almost darken the water as they rise to take breath'. He recalls dolphin pods shoaling and corralling fish in the creeks of the Thames. A documented army of

porbeagle, blue, mako and thresher sharks all followed
Britain's herring run each year. As for the white fish such as
cod, now increasingly scarce in our waters, the Royal
Commission of 1863 determined to uncover how much of
the precious herring cargo was being harvested each year by
these non-human predators. They estimated that over the
seven prime months of herring harvest, cod alone were
taking 29 *billion* herring. In addition, the Commission
estimated that the gannets of St Kilda were taking a further
214 million fish. In the nineteenth century, fishermen off
the Dogger Bank could catch one large cod every three
minutes. Common skate could weigh up to 90kg (200 lb).
Haddock, swimming past the coast of Yorkshire, formed
shoals that spanned 160km (100 miles). Such was the
abundance of British seas – an abundance that two centuries
of commercial fishing has wiped from our memories and
lives. Many of these large pelagic fish would have kept
mackerel stocks in check. Yet these, too, were far from the
biggest fish in the sea.

Their rightful predator, in turn, represented one of the
most impressive feats of nature's engineering on our planet.
With the oldest adults weighing up to 360kg (800lb), and
growing to over twice the length of a man, these impressive
predators, like super-sized mackerel, would slice through the
ocean at 40km per hour (22 knots). They have been
described by Sir David Attenborough as the 'ultimate fish'.
Serrated and sharpened to the last degree, their teardrop-
shaped bodies demonstrate hydro-dynamism unparalleled in
the animal kingdom. Retractable fins unfold during a tight
turn, aiding steering mid-pursuit, but are packed away
during a straight chase. Huge reserves of red muscle, designed
for endurance, mean this predator rarely tires on the hunt. In
every way, the Atlantic bluefin tuna is perfectly evolved to
visit trauma and destruction on the ocean's smaller fish.

In its early years, the bluefin undergoes one of nature's
most extreme transformations from pygmy to giant. It

commences life just three millimetres in length, easily
swallowed by a large number of predators: as a result, just
one in 40 bluefins will make it to adulthood. Like the
hungry caterpillar, as a bluefin tuna grows, so does its
appetite. It begins by predating microscopic copepods or
crustaceans, but by adulthood, a small group of bluefin can
reduce a swarm of mackerel to a glinting shower of scales in
less than an hour. Bluefin are powerful predators of squid,
and like any apex predator (being eaten, rarely, by only the
largest sharks or toothed whales) they regulate smaller
predators, from mackerel to haddock, that would otherwise
prey upon our sand eels.

Like most of the ocean's top predators, Atlantic bluefin
will sail through thousands of kilometres of marine desert
(as indeed much of the open ocean is), to find those
seamounts and upwellings where algae and plankton thrive.
In recent years, after four decades of absence, bluefin have
once again reached Britain's western waters – and the North
Sea. Here, so strong are the problems of continued
overfishing, that the true effect of tuna upon the ecosystem
is still hard to uncover. Given a chance, it has been shown
that predators like bluefin need only eat as little as 13 per
cent of small pelagic fish such as mackerel in an area, to
radically reduce the reproductive success of that species, and
as such, protect smaller fish from overly high rates of
predation in turn.

Given any chance to recover, Britain's bluefin tuna have
the potential to prove powerful puffin protectors, removing
the excess of piscivorous predators that vie with them for
sand eels each summer. For now, every bluefin ripped from
our waters denies our seabirds a fair feast. And the absence
of even one player, in its proper numbers, transforms life not
only in the ocean but also for those pelagic species who nest
along our coasts each summer on the land.

Just as the tiny lives of zooplankton grazers dictate the lives of seabirds, pelagic fish and super-predators like the Atlantic bluefin, so they also shape the fragile existence of Britain's beloved but endangered dolphins. Bottlenose dolphins, our largest and most often sighted species, need 33,000 calories a day – the equivalent of 60 salmon slices – in order to sustain their Olympian metabolisms. Like gannets, dolphins will feed intensively in certain areas, harvesting the fruits of the phytoplankton forests off our coasts. Britain plays host not only to a further five dolphin species, but also to a regular 21 species of shark. Each arrives with its own highly adapted diet in mind. Gentle, gaping basking sharks, the second-largest fish on Earth, arrive in our western waters each summer, adapted to sift enormous quantities of zooplankton. Blue sharks, a pelagic species far less familiar to many, feast on squid, octopus, shrimp and crab. The shortfin mako, found in our warmest waters, takes things up a level; its 25mm razor teeth are adapted to shred tuna and even small seals. Known to few, the mako is the second most powerful predator of British waters.

At the very height of the predatory ladder swims a hunter unmatched by any – invulnerable to all except ourselves, and perhaps the odd male narwhal possessed of a long tusk and a great deal of good luck. Orca – formerly better known as killer whales, are the world's largest dolphins, and the marine equivalent of African painted dogs: pack predators so intelligent and sophisticated that they win the game almost every time. Their lives, too, flow from the zooplankton forests around our coasts. Shetland's resident and migratory orca feed primarily on seals. Migratory orca, arriving from Scandinavia into the North Sea each autumn, often hunt mackerel shoals, their enormous brains now trained to lock onto and follow trawlers. By feasting on mesopredators, dolphins, orca and our larger piscivorous sharks protect the smallest grazers of the ocean, removing their predators while rarely stooping to swallow a meal as small as a mouthful of zooplankton.

Yet in drowning dolphins as by-catch in nets, filling their stomachs with plastic, and their bloodstreams with polychlorinated biphenyls – a dangerous ocean pollutant that reduces the reproductive rates of dolphins and orca – we ceaselessly hamper the efforts of these masterful beasts to regulate the seas beside us. Of all British waters, only 0.01 per cent is free from the effects of industrialised fishing, and our numbers of orca, dolphins and larger sharks are mere relics of what they once were – or could be, again, in the future. Were such animals present in numbers that our seas are supposed to sustain, the mackerel or haddock that prey upon sand eels would themselves be naturally reduced. Under such conditions, the plight of our most imperilled seabirds, from our puffins to our kittiwakes, would itself be greatly helped.

None of these larger predators, however, may be enough to stop the growing silence in our seas. As our oceans warm, it is predicted that phytoplankton will suffer – for the reason that warmer waters contain less oxygen, reducing their ability to photosynthesise. Since 1950, over 40 per cent of the world's phytoplankton garden has already wilted and died. Phytoplankton, however, are sentient to the changes in their environment. As a result, many communities in the North Atlantic are predicted to move north into cooler, Arctic waters. This represents quite simply one of the largest and most frightening shifts of biomass on Earth – a migration of prey away from the fish, seabird and cetacean predators that have evolved to find it. And already, sand eels are showing us what future may lie ahead.

In 2004, conservationists collected samples of sand eels that had been dropped by puffins and other seabirds in their colonies. The sand eels were found to be far lower in energy than expected. Between 2000 and 2004, only half of first-year sand eels are thought to have grown to maturity, compared to 80 per cent in normal years. As sand eels do not feed in winter, they are reliant on the energy reserves snatched during a short summer banquet: as few as 10 per cent will survive a normal winter. The length of a sand eel, in a particular year, is a good determinant of how many planktonic calories it has piled on in the preceding summer. In 2004, in particular, it was found that diatom densities from within recovered sand eels were critically low, stunting their growth rate. When phytoplankton decline, starvation creeps like a cancer through every level of the ocean. There are, however, one group of animals on which even phytoplankton depend: animals that not only live within the ocean, but also create ecosystems from scratch. These are the largest animals ever to have lived: the great whales.

In the ocean of a thousand years ago, Britain's seas were governed by giants. Millennia before, hunter-gatherers and early farming settlers had made short work of our elephants, mammoths and rhinos, and then of the wild horses, aurochs and larger predators onshore. But it would only be far later that our land-based species would tackle the giants of the seas. Indeed, our once-familiar whale neighbours were denoted by the Old English word, *hwael*, being most likely abundant and visible, too, from most British shores.

Twelve-metre-long Atlantic grey whales, now lost entirely from the world, are thought to have swum northwards from Mediterranean calving grounds to feast on ocean shrimp around Britain's coasts. The grey whale is a masterful swimmer in shallow waters, bears its young in bays, and rarely runs aground. So how many of these coastal whales did Britain's coastline once harbour? In California, one localised population, now thought to have attained carrying capacity, numbers around 26,000 animals. A millennium ago, similar or greater numbers may have graced our waters too, prior to their eradication by the end of the seventeenth century. Of all the lost sights in the British Isles, the thought that we once had a thriving population of friendly coastal whales, gracing our bays and inlets, is at once indescribably wondrous – and sad.

The North Atlantic right whale, growing to 18m in length, was, at one time, perhaps the commonest of all whales around our shores. These docile whales were 'right' for the early whalers, as they dwelt largely on the surface. Once killed, they would float, due to their high content of blubber – making them the perfect harvest for whale oil. We know from the sheer archaeological frequency with which whales were butchered on-shore, well before the time of commercial whaling, quite how many once washed up naturally on our shores. And rather like the dead badgers on our roadsides, the visible mortality of any species is often a mere hint of the abundance of its living peers. Both grey and right whales suffered the earliest declines, being so readily killed close to our coastline. As late as the eighteenth century, however, giant pelagic whales, which we now associate with deep offshore waters, were once common sights from British shores as well.

Goldsmith recalled the arrival of the herring run off the British west coast: 'The whole water seems alive; and is seen so black with them to a great distance, that the numbers seem inexhaustible.' He adds that so great was the spectacle, it would each year 'alter the very appearance of the ocean'. He described 'fin-fish' (fin whales) and 'the cachelot' (sperm whale) plundering rafts of herring covering several kilometres. Due to the nature of Britain's marine channels – the southern North Sea and eastern English Channel forming a shallow bottleneck, as the whale swims – it seems likely that Britain's whale abundance would always have been greatest off our western coast. However, there is no such shallow water to the north of our island, so pelagic giants like fin and sperm whales would also have been able to pass unhindered round the northern tip of Britain – and into the North Sea.

It is shocking indeed to find that even until the early twentieth century, the greatest animals ever to have lived may, on a regular basis, have called British waters home. Between as recently as 1903 and 1914, and then, after the First World War, between 1920 and 1929, no fewer than 85 *blue* whales were taken by the Shetland whaling industry. The same records suggest that northern right whales and humpback whales also persisted in numbers here far later than elsewhere, which correlates not only with Shetland's remote situation, but also with the quality of its waters for our vestigial whale populations even to this day. It is strange indeed to think that while most of the British public are, to some degree, aware that wolves and bears once walked our land, most have no idea that until a century ago, 30-metre giants, larger than dinosaurs, would have cruised our northern and western waters.

Only by looking backwards, through the bloody annals of whaling, can we understand quite how common great whales once were. Globally, it is thought that whaling robbed three *million* whales from our planet between 1900 and 1999. But whaling commenced far earlier than this: in the eleventh century, under the Basques. There would have been millions *more* whales than this in our ocean a thousand years ago.

In the North Atlantic, studies of genetic diversity patterns suggest that before the commencement of whaling, our waters were home, in recent times, to 240,000 humpback, 360,000 fin and 265,000 minke whales. These, it must be noted, are only the species where enough whales still exist to make such genealogical back-dating possible. No firm estimates have ever been made of the sperm whales, blue whales, right whales and grey whales that once swam in our seas.

In all, adding in even the most conservative estimates of other whale species suggests that more than a *million* whales would once have swum in the ocean that lashes

Britain's west coast. And these common giants would have transformed our oceans.

Whilst the lives of almost all other species in the ocean are dictated from the bottom up – from the abundance of plankton in our ocean – the great whales are different. Whales *fertilise* the ocean – shifting nutrients from deep to shallow waters. As they feed at great depths, the pressures of the deep shut down many of their bodily functions, so whales only defecate at the surface of the ocean. In doing so, they expel huge quantities of faeces rich in iron – expulsions that are ten *million* times richer than the surrounding ocean. Studies at the University of Tasmania have shown that plankton prospers best when fertilised by the iron of whale faeces. Further studies, in the Gulf of Maine, have calculated that prior to their desecration by whalers, the great whales would have released three times the nitrogen – a fertilising agent – into the water as was naturally sequestered from the atmosphere. In undiminished numbers, whales create the very gardens in which phytoplankton grow, enhancing the fertility of the ocean and thereby creating the basis of entire plankton-based trophic cascades.

Whales do not simply feed within the ocean – they create and *maintain* populations of their prey. In 2014, a remarkable paper, entitled 'Whales sustain fisheries,' was published in the respected journal, *Marine Mammal Science*. The authors had set out to examine the popular assertion that whales would, logically, compete against human fisheries, by harvesting large quantities of marine resources, which might otherwise be harvested by us. They also examined the hypothesis that with a decline in blue whales, krill – the whale's food – would increase. In fact, both hypotheses turned out to be completely wrong.

Blue whales in the Southern Ocean were, instead, supercharging the primary productivity of the ocean, by defecating iron-rich faeces at the surface. The scientists calculated that the amount of iron defecated each year would lead to primary productivity – namely, the formation of new phytoplankton blooms, and the krill that fed on them – of sufficient volume to support the blue whale population here. In other words, the whales were *creating* as much food as they were eating. In addition, whales were also creating entire ecosystems, through changing the very composition of the ocean. And therefore any attempt to remove whales, it was concluded, would simply reduce, not enhance, a fisherman's catch – as whales provide the very basis for fisheries to thrive.

Whales, being unique in their physical size, are capable of vectoring huge quantities of iron into the photic zone – the upper layer of the ocean where sunlight permeates and thus photosynthesis can occur; leading to the formation of phytoplankton. For this, both sunlight and iron are required. To a lesser degree, seals – also feeding at depth, and defecating at the surface – achieve the same end, but not to anywhere near the same degree as the great whales. And in those many areas where whales and seals have been most heavily hunted, declines in phytoplankton – the basis of the marine food chain – have been greatest. Without phytoplankton and the zooplankton grazing it, the basis for much marine life – from sand eels to mackerel, all the way to bluefin, dolphins and orca – becomes far less viable. In other words, without whales, in undiminished numbers, the ocean's very productivity is put at risk – and has indeed been this way now for centuries. And for the ocean to be well-fertilised with iron, we need not just a handful of whales, but the millions that swam in our oceans three centuries ago.

Whilst the impact of pelagic whales – deeper-water species such as the fin and sperm whale – would once have been enormous in British waters, in light of what whales do,

it is possible that the complete eradication of grey whales –
our coastline giants – may have been amongst the greatest
losses of any single cornerstone species from our shores.
When we consider that entire populations of these gentle
giants would once have moved seasonally along our
shoreline, creating the conditions for marine life to thrive, it
is hard to imagine what our coasts have lost in abundance,
and diversity, as the result of their absence.

The ramifications of whale loss extend far beyond the
mere creation of vast marine habitats. The environmentalist
George Monbiot notes that even the California condor, a
predominantly coastal mega-scavenger, was historically
recorded as the key scavenger of whales washed up on the
American shoreline. Bereft of whales, the condors have
targeted deceased terrestrial mammals, often ingesting lead
shot in the process. Across the world, and here in the UK,
the impacts of losing giant whales are more far-reaching
than we might think.

In the context of Britain, and its seas, the apparently
unrelated lives of the puffin, the sand eel and the
phytoplankton communities around Britain's coasts may
require the return of giant whales, in enormous numbers, if
they are to survive – strange and alien as that idea may seem.
And every whale around our coastline becomes an asset not
only to seabird life, but also to the fight against climate
change itself, as they create the conditions for ocean
photosynthesis. Indeed, it is hypothesised that prior to
whaling, whales would have played a significant role in
mitigating climate change; providing the conditions by
which diatoms, and other phytoplankton, lock away carbon
dioxide in the ocean.

What is promising is how whales are beginning to stray
once more into our seas; seas they once called home. The
UK Cetacean Strandings Investigation Programme now
reports more and more whales washing up on our shores.
For the first time since the eighteenth century, humpback

whales have been sighted not only around Shetland but also
off the coast of Norfolk. North Sea sperm whales have
washed up too. Between 2011 and 2017, 4,896 whales,
dolphins and porpoises washed up on British beaches; up 15
per cent on the previous seven years. Grim as these deaths
may be, they hint strongly that offshore, many more whales
may now be singing in our seas. Simply by ceding right of
way, we can allow animals as large and mighty as the fin
whale – eight times the weight of an elephant, and longer-
lived – back into our lives. As our seas founder on the brink
of starvation, only the greatest of living animals have the
power to protect the very smallest. Through that partnership,
alone, will the creatures of the sea endure.

In August 2012, I was privileged to witness the true majesty
of a teeming, whale-governed ocean for myself – albeit far
from our own depleted waters here at home. I had travelled
to Kodiak Island, in Alaska, as part of a filming expedition.
On arrival at Kodiak harbour, I was staggered to find that
the piscivorous bald eagle came not in its tens, but hundreds,
as white-tailed eagles would once have done on the lakes,
wetlands and coastlines of our own island. The river creeks
were lithe and frothing with salmon, of several species, all
making their way inland to spawn. Then, one calm dawn,
we set out into Kodiak Sound in search of whales. As we
did, I saw something I have never witnessed off the British
coast; as the boat left Kodiak harbour, droves of pelagic fish
leaped out of the water, clearing the bows of the boat at
short range in slippery panic.

As we headed out to sea, there were black rafts of 'sea
parrots' – tufted and horned puffins – as far as the eye could
see. The seabird colonies off the coast of Alaska are the
largest in the world; dwarfing those of the Falklands, and
our own, and consisting in all of more than fifty million

nesting birds. A wide array of swimming auks and diving terns peppered the water, and the frothing fish gave it form. It became startling to me, as we sailed, and bald eagles sometimes passed by to skim fish from the water, that an ocean could be so disrupted with living forms. Then, the first humpback plume shattered the water.

Being born in Britain, and raised upon the meagre standards of our nation's wildlife, it had occurred to me that we were in search of *a whale*. This, after all, was where my expectations had been set. And yet, within the next hour, we would, without any real effort, encounter more than 50 humpback whales. The sea was broken not only with their plumes but also with their spectacular breaches, as they lifted barnacle-encrusted bodies, each weighing 30,000kg, almost clean out of the water, the spray from their splashes carrying through the air onto the deck. All the while, the waters around were a constant feeding frenzy of gulls, terns, guillemots and puffins. It would only occur to me on writing this book that this spectacle would have been no different from taking a small boat out from the coast of Devon, Pembrokeshire or Argyll three centuries ago. Since 1966, from just 1,400 individuals, the Alaskan population of humpback whales alone has increased to over 21,000. At this level, their local ecosystem impact has become profound, as they enrich Alaska's coastal waters *en masse*, vectoring iron into the photic zone and creating the basis from which plankton, fish, seabirds, eagles, otters and orca can come to thrive.

One day, if great whales return to British waters, in the numbers they once held, we may all gaze out on the azure orange calm of the western Atlantic as the distant plumes of 'fin-fish' and the 'cachelot' catch the last rays of the evening sun. We will know that out there, beyond land, beyond us, schools of fish several kilometres long are locked in a deadly game of hide and seek with the largest animals on Earth. And, bearing sea-fish in their talons, eagles the size of barn doors will drift ashore to fertilise the land.

CHAPTER FIVE

Bees

If you were to visit China's Sichuan Province in spring, you would be met with an odd sight. Delicately balanced atop tall ladders, placed amid a sea of white flowers, artists with extra-long paint brushes appear to polish pear blossoms. Meticulously cleaning every single flower, whilst delicately reaching up to the outermost branches, requires intense concentration and poise. Yet those doing the painting are no artists. And this act isn't a labour of love, but born of agricultural necessity.

For millennia, human farmers have relied on a mutual contract with the natural world. In return for a bountiful local supply of pollen and nectar, from a host of different crop species, bees and other invertebrates provided a free pollination service – ensuring healthy returns later in the season for farmers. Over the course of human farming history, this contract has arguably been one of the most lucrative and important in the world, which according to Forbes is now worth over 500 *billion* dollars every year.

In recent decades, however, humans have tried to cheat on the contract. After all, alongside free pollination, nature has also always taken a cut of the spoils. With the invention of pesticides, humans decided that they weren't interested in sharing the percentage of crop-land lost to various moths, beetles and flies. And yet, by trying to cheat the system, we fundamentally broke it.

So intensive has the use of pesticides become in Sichuan's crop-producing region of south-western China, that insect life has all but disappeared. White fruit blossom paints the canvas of this otherwise bleak landscape, but the air no longer hums. The free pollination service once

provided by hordes of industrious workers has instead
enchained humans to the backbreaking work of manual
labour. Carrying a jar of freshly harvested pollen, hung on
strings around their necks, and armed with five foot
pollinating sticks (adorned with chicken feathers), Chinese
labourers now set out every day of the blossoming season
to ensure that their farms have a crop later in the year. It
is a sorry sight.

Classical ecological dogma describes ecosystems as top-
down pyramids; regulated by the most powerful apex
predators at the top, with a sprawling triangle of life
functioning below. But reimagine that pyramid as a spider
web, and you soon realise how cutting *any* of the strands can
have just as great an impact as removing those trophic
champions sitting at the top.

In fact, as seen in Sichuan, removing the smallest
components of nature can have the most devastating impacts.
But sadly, we cannot judge Sichuan. Indeed, the British have
sadly become world leaders in bee removal, too. Sterilised
through the removal of their herbs and plants; choked,
disoriented and killed by neonicotinoid sprays, bees are
faring worse in Britain than in most European countries. If
we want to restore thriving grasslands, diverse flora and the
basic ability to feed ourselves in the future – we must restore
the bee machine.

Of all the thousands of bee species around the world, there
is one that has been more successful, one on which we
more completely rely, than all others – a cornerstone for
our own existence: the western honeybee. As you read this,

millions of worker honeybees will be out collecting pollen, nectar and water, fulfilling a genetic urge that ensures that, from the moment they hatch, the bees work themselves at such an extraordinary rate that one day they simply fall out of the sky and die of exhaustion. Capable of pollinating the majority of known crop flowers, and critical for the pollination of such foods as almonds, squash, watermelons and many other staple fruit and vegetables, the efforts of these aerial labourers would be sorely missed if they were to disappear. But how did the honeybee rise to become such a workaholic?

A typical honeybee hive can contain upwards of 50,000 individuals. The largest super-hives, which occur naturally in suitable cavities, can host quadruple that. If we attempted to live at such extreme population densities, the result would quickly spiral into catastrophic failure. The secret to the successes of a honeybee hive are manifold, but three things stand out: language, order, and luck.

From the moment an egg is laid by a queen bee, it receives some of the best midwifery in the world. Tended to by young worker bees, the egg is kept at just the right temperature, and just the right humidity, to hatch after exactly three days. The larva has no eyes, no legs and no antennae. At this stage, it is simply a tiny vacuum cleaner – devouring the honey and royal jelly fed to it by older siblings. On the 21st day, what started off life as a grub the size of a grain of rice emerges from the pupa it formed inside its hexagonal cell as a fully formed bee. Extraordinarily, it emerges like a pre-programmed smart-phone, fully coded to carry out the functions it will stick to for the rest of its working existence.

The bee brain measures just a single cubic millimetre. Packed within that are a million neurons: a tiny super-computer that produces one of the most beautiful languages yet discovered. Famously known as the 'waggle dance' – this vibratory, olfactory and auditory performance is completed entirely in the dark of the hive interior – yet allows a worker bee who has found a good pollen supply to tell its comrades exactly how far and in what direction they must fly to find the same.

It is more than evident when watching a healthy honeybee hive on a spring morning – as they funnel out and all fly off, making a literal bee-line – that they know *exactly* where they are going. The returning bees, so laden with willow, hazel and blackthorn pollen they are barely still airborne, are testament to the success of the discussions being carried out between workers within the hive. But what do the worker bees get in return for all their hard work?

Because it is only the single queen in the hive who lays eggs, the workers themselves never get a chance to reproduce. Once again, it comes down to genetics. The worker bees are all female, and all siblings. Therefore, although they don't get a chance to pass their genes on through direct reproduction, by helping the hive raise their sisters, a few of which will become new princesses, they are helping pass on their shared genes via this route. And due to a quirk in eusocial insect genetics – they actually on average share three-quarters of their genes with these sister workers and future princesses. Humans only share half of their genes with their progeny. Eusociality has arisen in many different insect species across the world and also in one mammal – the remarkably ugly naked mole-rat of sub-Saharan Africa. Of course, the individual bees don't 'know' about their genetic relationships – so why don't the workers revolt?

It transpires that this whole society is kept in check by the pheromonal offerings of one bee – the queen. Researchers in New Zealand discovered that a single chemical – homovanillyl alcohol, or HVA – prevents worker bees from developing aversions, helping keep any tendency to revolt in check. Interestingly, HVA is chemically similar to dopamine – the chemical in human bodies that contributes to the feeling of pleasure. This is just one of many hundreds of chemicals working alongside one another to keep the delicate balance of honeybee hierarchy in check. If the queen is weakened, or removed entirely, the whole hive descends into anarchy. If a single bee from a different hive finds herself in amongst the wrong comb, the workers soon determine that something doesn't smell right, and she is savagely evicted, or worse, simply pulled apart by the defensive hive owners. With thousands of man-made chemicals filtering into the natural world as aerosols, in our water and into soils, the true effects of their interaction with such complex systems as honeybee hierarchy are still barely understood. Safe to say, the effects are likely to be increasingly negative.

The final component to honeybee success comes down to a little luck. Honeycomb must be one of the most instantly recognised natural materials. And it is a brilliantly efficient way of storing honey and pollen – the ingredients that allow honeybees to feed during the winter when the ground is frozen and there is no pollen for many months at a time. Most social bee species overwinter as individual mated queens, which then have to go through the laborious task of starting a new colony come the spring. Many fail. Yet by overwintering with a massive food supply, honeybees have a huge head start as they come out of the winter. With thousands of workers deployed to harvest the early pollen as it becomes available, they rapidly grow their colony in early spring and are soon able to put out their first of many swarms. Without honeycomb, that simply wouldn't be possible.

The mystery of the exquisitely uniform hexagonal cells has fascinated humans for millennia. The Roman philosopher Marcus Terentius Varro proposed in the first century BC that honeybees build their cells in order to achieve the best economy of material. Charles Darwin further theorised that a swarm that could secrete wax using the least honey would be most successful. But it turns out, the hexagonal structure of honeycomb isn't perfect, and could actually be down to chance physics.

Most bee species across the world build some sort of cup to store nectar. Bumblebees, veritable giants in the bee world, build large cells from wax to store their food. The early ancestors of honeybees would have used a similar method of food storage during their breeding season over spring and summer. By packing enough of these irregular cylinders into a small space, such that their cell walls touch and fuse, the result is a comb of polygons. Packed together, with walls forced to fuse, the result is a continuous hexagonal lattice. Over time, this has been refined as honeybees evolved and the final result is one of the most exquisite pieces of art in the natural world: honeycomb. This same method of cell construction has independently evolved in paper wasps, who also use a hexagonal matrix to raise their larvae in.

It is a combination of those three things: the efficiency of their language, the intensity of their order, and the latticed shape of their comb structure, that has allowed honeybees to become such productive workaholics and in turn, become our most valuable ally for global food production.

Whilst honeybees may be the economic superstars of the bee world, bumblebees are arguably the most loved, as well as invaluable pollinators of a range of plants. With their stripy suits, gentle buzz and endearing antennae, they have inspired generations of children's authors, while their image helps sell everything from hand soap to holidays to honey.

For a temperate-thriving species, a warm furry coat is an excellent idea, which even allows the aptly named *Bombus polaris* to thrive well inside the Artic Circle. However, the downside of this additional insulation is that bumblebees rapidly overheat. Few bumblebees live in the Mediterranean and almost none persist in the tropics. At 44°C, bumblebees become so overheated inside their extra layers that they die. With climate breakdown advancing global temperatures every year, over the coming decades, our various European bumblebee species are going to be pushed further north on a thermocline of heat extinction. This could be bad news for the many different flora species dependent on their pollination service.

In 1859, Charles Darwin published an observation in his renowned work *On the Origin of Species by Means of Natural Selection*. It was one that soon came back to haunt him. Residing at Down House, now on the southern outskirts of London, Darwin performed a number of experiments in the local meadows to determine the importance of bee pollination.

Using a fine woven mesh net, he built a series of cuboid cages and placed them over patches of red clover as they flowered, whilst leaving other patches open to invertebrates. His results were utterly conclusive: the uncovered clover produced a healthy seed crop, yet the covered crop didn't produce a single seed. Darwin wrote 'I have very little doubt, that if the whole genus of humble-bees became extinct or very rare in England, the heartsease and red clover would become very rare or wholly disappear.' By Darwin's era, this result wasn't a huge surprise – the relationship between invertebrate pollinators and flowers was beginning to be understood, but it was Darwin's further extrapolation of the impacts of this that stood out. He continued:

> The number of humble-bees in any district depends in great degree on the number of field-mice which destroy their combs and nests ... Now the number of mice is largely dependent as everyone knows on the number of cats ... Hence it is quite credible that the presence of a feline animal in large numbers in a district might deter-mine through the intervention first of mice and then of bees, the frequency of certain flower in that district!

Bees need flowers, as much as flowers need bees. If the pollinating species of bee disappear, the knock-on could ripple through the food web. This is the earliest known example of one of the main staples of modern ecology – a food chain. And it is one that was occurring long before Darwin wrote about it. In meadows and woodlands across Europe, wildcats once provided nature's finest rodent removal service.

Under barely audible footsteps, these highly tuned killing machines prowl the margins. Every so often, they freeze. Poised, the only movement comes from one ear, deftly rotating to pinpoint the sound of tiny footsteps on wet grass. A slight head-turn tracks the sound as it moves between a tunnel system at the base of the grass. Time ticks, but the wildcat doesn't move. Every muscle in its body is preparing its next move. The second ear joins the first, both now pointing directly forward – two tiny satellite dishes picking out every hint of sound and converting it into a 3D movement map. In slow motion, the wildcat lifts one front paw, twitches its back feet, then catapults forward with both front feet outstretched. Pinned to the deck, one more rodent is swiftly snuffed out.

In spring, this creates just the right conditions that a newly emerged bumblebee queen is looking for. An empty vole burrow, with a cosy nest of moss and grasses safely hidden at the end of an underground tunnel, is the best place for a bumblebee queen to raise her own family. And so the food web continues. Bees need vole holes. Voles need clover. Clover needs bees. And wildcats help keep the whole system in balance.

Unfortunately for the ever self-critical Darwin, his early theorising turned out to be a little wide of the mark. On learning, after publication of his famous book, that red clover is actually capable of being pollinated by a few different species of bee and therefore a little safer from extinction should its main pollinator disappear, he wrote to his neighbour angrily berating his oversight: 'I beg a million pardons. Abuse me to any degree but forgive me – it is all an illusion (but almost excusable) about the Bees. I do so hope that you have not wasted anytime for my stupid blunder. I hate myself, I hate clover and I hate bees.'

The variation in size, colour and shape of flowers is extraordinary. From the pinhead-sized *Wolffia globosa*, the world's smallest flowering plant, to *Rafflesia arnoldii,* which produces massive red flowers a metre in diameter and up to 11kg in weight – there are nearly half a million flowering plant species, of every conceivable configuration, in between. Almost all of these have co-evolved with at least one insect pollinator, an arms race in which the flower offers up nectar in reward for pollination and the insects (plus a few birds and the odd mammal) come up with increasingly efficient ways to harvest that nectar. Some nectar-drinkers have learnt to bypass the pollination mechanism entirely, instead drilling a hole through the side of the flower and sipping straight from the nectary. Others use clever electronic cues to determine whether or not a flower has recently been visited by another pollinator – and thus move on rather than expend their energy at an empty bar. Some bumblebee species, particularly when pollen-deprived, have learnt to bite flower leaves to induce their buds to flower sooner. But a discovery in 2018 forever changed the way that we view the ever-evolving relationship between bees and flowers.

The natural world is full of sounds. Whilst some – such as leaves rustling or wind blowing – are easily filtered out and ignored by us, others – such as a bumblebee's buzz – are distinctive, cutting through the background soundscape to be readily detected by our ears. But it turns out ours are not the only ears processing this sound.

Enter the evening primrose. Producing tall stalks, sometimes well over a metre tall, covered in many ten-pence-sized yellow flowers in the shape of shallow satellite dishes, this garden favourite holds a remarkable secret. When Tel Aviv University researcher Lilach Hadany began to play plants sounds to see how they responded, she discovered that evening primroses *have ears*. When played white noise, high frequency sounds or no noise, the plants seemed to carry on as normal. But when played bee sounds, or low frequency sounds, the response was unmistakeable. Within three minutes of being exposed to these, the sugar concentration in the plants' nectar increased from an average of 14.5 per cent to 20 per cent. The flowers could apparently *hear* the presence of bees and responded by upping their sugary offering. The research is still in progress, but if the pollinators are held at such flowers for longer, increasing the chances of successful cross pollination – it makes sense for the flower to provide this richer reward.

Excitingly, Hadany's team also ran tests to see how the shape of a flower would respond to different vibrations. Sure enough, evening primrose flowers resonate at pollinator frequencies. As Hadany explains, 'It's important for them to be able to sense their environment – especially if they cannot go anywhere.' It is extremely likely that this new field of study, dubbed phytoacoustics, will yield many more previously unknown relationships between plants and pollinators. In the decades to come, this line of research is likely to only further cement the crucial role that such species play in their ecosystems.

Our social bees are important. They have beautifully complex lives, and ruthlessly efficient societies. Yet every day throughout spring, summer and autumn, bees are being pincered, skewered and plucked from the air. Although they are often armed with a powerful sting, extraordinary reflexes and the power of flight, the natural world certainly doesn't disappoint when it comes to the army of predators that specialise in the delicate business of bee removal. As a result, social bees are not only a keystone species for the services they provide – but also for the menu that they serve to an array of bee-eating specialists.

If one of our many bee species manages to make it through the suite of parasites that invade their nests, the cuckoo bees that evict their siblings, and the hornets that wait like the school bully outside their nest entrance – they have a chance to set off and find food. But the flowers that they visit can hold a nasty surprise. Occasionally, without warning, the petals of a tasty flower appear to snap shut, trapping a visiting bee which is then dragged down between the petals, never to be seen again. The culprit is not the plant, but a spider.

With its bulbous abdomen, four pairs of muscular legs and eight eyes, a *Misumena vatia* crab spider is about the size of a common garden spider. But unlike their cousins, these colourful predators don't rely on a web to catch their prey. Instead, they rely on a brilliant piece of biology. Thanks to specialised pigment cells, these arachnid chameleons are able to completely change their colour to match the petals of the plant they lie in wait on. One week they might be white, nestled around the rim of the ox-eye daisies. Two weeks later, they might be yellow, colour-matching the meadow buttercups. A month on, the very same spider might be pale cream, a spider swatch camouflaged against the flowers of hogweed. Sitting patiently with its front legs held open like a pair of pincers – hence the name crab spider – it waits for its next victim to land. As soon as the prey buzzes within range, the pincers clamp shut, the jaws deliver a venomous bite, and the bee is disarmed. As the spider's digestive enzymes liquify a million bee cells into protein soup, its fangs work like straws to suck the liquid from its victim.

Should the bee manage to avoid these snap traps hidden in their favourite food, they then have to fly the avian predator gauntlet to get their cargo back to the colony. In mature wood pastures, spotted flycatchers wait on branch outposts, ready to launch aerial assaults on pollen-laden bumblebees. Plucked from the air, the bees are unceremoniously snuffed out, their stinger rubbed against the branch until rendered harmless. In scrubby patches of woodland, shrikes specialise in skewering bees on blackthorn, often alongside a cache of other unfortunate victims including lizards, songbirds and rodents. Now almost extinct in Britain, red-backed shrikes once thrived on our bee biomass.

But of all the species capable of ending a bee's day, there is one that is better than all others. Graceful, elegant and swift, with a fluting song that sounds like a warm Mediterranean evening and a colour palette that rivals the flashiest Amazonian parrot, the bee-eater is comically cute. A bright yellow bib, a turquoise chest, whisps of blue in the cheeks and a jet-black bandit mask adorn its front. A deep graduated ochre paints its back, leading down to a long and tapering green tail. In the right light, this exquisite plumage shimmers. But good looks can be deceiving. The black pupils in its crimson eyes miss

nothing. And its gently curved long bill is a weapon
trained to extinguish bees. These birds are rare visitors to
our shores, but with warmer summers, the appearance of
a few pairs seems to be more frequent, with breeding
recorded more times here in the past decade than the
whole of the twentieth century. Perhaps one day,
showcasing these hungry avian superstars might help fly
the flag for the precipitous invertebrate decline we are
currently witnessing amongst our social bees.

So our honeybees are economically essential pollinators
with incredibly complex lives and our bumblebees are
aerial protein parcels being *listened* to by flowers, but
we have barely scratched the surface when it comes to
the extraordinary diversity within the wider bee world.
In Britain, over 90 per cent of our bee species are neither
bumblebees nor honeybees (of which there is just a single
European species), but solitary bees. Rarely noticed
yet found from Scilly to Shetland, the solutions that
these underappreciated creatures have come up with to
fit into their natural worlds are both extraordinary and
outstanding.

There is a species of leafcutter bee that neatly chisels
precise circles into rose leaves, taking up to 40 cut-out discs
back as wallpaper for each of her nest chambers. One of our
smallest species of British bee likes to bury itself in the head
of dandelions, appearing many minutes later completely
coated in yellow pollen. There are bees that nest in sand,
mud and wood. Cuckoo bees – as their name suggests –
attempt to exploit the work of others. Miner bees mine,
digger bees dig, mason bees nest in wall holes, and wool
carder bees strip the hair from furry leaves and whip it up
into a nest lining. But it gets better.

Our smallest bee is so tiny that it nests in the holes left
abandoned by woodworms. At just 6mm long, the jet-black
small scissor bee is a very particular customer when looking
for a home for its eggs. But not as picky as one of its
relatives. The metallic-coloured blue carpenter bee digs out
the ends of broken bramble stems. But this only works with
the help of giants – for they aren't capable of cracking the
bramble stems themselves. A large herbivore shattering dried
bramble stems in pursuit of a new feeding opportunity or
bedding area reveals exactly what the blue carpenter bee
needs – the exposed bramble pith. Carefully boring a hole
down into this natural insulation – these solitary bees tuck
themselves in for the winter – it hibernates in a well-
protected spiky fort and likely hopes that it isn't awoken by
the ground-tremoring footsteps of one of the giants that
helped create its slumber-shelter.

Though adorable, there is another solitary bee that wins
the award for ingenuity and endurance. The gold-fringed
mason bee is as pretty as it sounds – with each leg and
abdominal segment embroidered with a fine plume of
gold hairs. Each spring, female gold-fringed mason bees set
out on a mission, searching for a place to lay their eggs.
Flying low over herb-rich grasslands, they frequently dive
into the flora to inspect possible sites. But unlike the
miners, diggers or masons, these discerning, gilded bees are
after a ready-made apartment block. Following one of
these bees, and checking what they're after, you'll likely be
surprised to see they are meticulously inspecting every
snail shell they come across. An empty Roman snail shell is
discarded – too big. A young brown-lipped snail shell is
quickly ignored – too small. Like a tiny Goldilocks testing
bear porridge, this shell-shopper is particularly keen on
vacant garden snail shells. Once decided, the female bee
will lay a few eggs in neatly constructed chambers within
the spiral structure of the shell, before plugging up the
door with a mastic made from chewed leaf pulp. In a final

act of maternal care, she plasters the shell in a fine coat of leaf mastic, a sort of camouflage jacket with a purpose still yet to be discovered.

In 1962, Rachel Carson predicted the devastating effects that a tidal wave of synthetic pesticides sweeping the world would have in her visionary book *Silent Spring*. She has been proven emphatically right. DDT might be the most notorious insecticide, but some modern neonicotinoids are *ten thousand times* more toxic. Every year, great swathes of the countryside are coated in a blanket of invertebrate death as sprayers target everything from potatoes deep underground to apple buds waiting to fruit. Time and again, humans continue to try and cheat the natural world, reaping it of its bounty in dangerously over-stretched ecosystems, whilst expecting the whole process to continue without fail. By breaking our natural contracts, we are breaking these systems. Many are now in freefall: some, as in Sichuan, have collapsed. Why should the natural world return free pollination when it is being annihilated by insecticides, herbicides and fungicides? Our global farming systems have become so efficient at doing one thing, they appear to have forgotten that they simply cannot function without natural pollinators.

Bees *can* bounce back. Like many insect species, their productivity when left to their own devices is exceptional. But this only works at a landscape level. Individual wildlife-friendly farmers are leading the charge but are too often let down by neighbours that insist on using powerful insecticides. If we can repair this strand in the food web, we can reverse the destruction of natural capital occurring all around us.

In June 2018, I awoke in the Carpathian Mountains of eastern Europe to the uncomfortable realisation that I had booked my rural retreat far too close to a motorway or at least, a major road. Sleepily coming to and parting the curtains, then dressing, I struggled to define where the dull, persistent roar was emanating from. Stepping outside, it became quickly apparent that the culprit was not a motorway – but honeybees.

There are few, if any, landscapes left in the UK that reverberate so loudly with pollinating bees that the sound wakes people up in the morning. Whilst the 'bee-loud glade' referenced in W. B. Yeats's poetry is frequently remarked upon, few of us have heard bees as they are supposed to be heard: a purposeful roar, loud enough to get you out of bed.

We are blessed, however, that right now, large tracts of Europe, especially those in eastern countries such as Latvia, Estonia, Poland and Belarus, and those areas dominated by the Carpathian's mosaic of forests and meadows, such as Slovakia, northern Hungary, Romania and Bulgaria, have yet to fall silent. These landscapes abound not only with bees – but with all that bees create.

In these sympathetically farmed landscapes, where small herds, small crops and zero pesticides or herbicides still conspire to provide enough food for local people, it is not only bees that are common. Turtle doves, on the verge of extinction in the UK, grub seeds from the abundant chickweed, knotweed and fumitory pollinated by honeybees that can number five million within the hives of a single village. It might be easy for us in western countries, with higher populations to feed, to think that such methods are outdated, but they will, in all likelihood, last far longer into the future.

The return of honeybees to our landscapes, in the vast numbers that were once normal, will signal the return of a phenomenon far more important for the recovery of nature than biodiversity: *bio-abundance*. And the actions of a society of millions, writ large across our landscapes, can, in a very short space of time, rejuvenate those landscapes once again. But as Sichuan reminds us, there is another future. If we continue to ignore the warning signs, we too will soon have to listen to the ever greater silence in our land – and suffer the consequences of our broken contracts.

CHAPTER SIX

Cattle and Horses

In early May 2016, on the steppes of the Hustai National Park in Mongolia, I had been reliably told the weather would be sunny, calm and warm. Instead, a blizzard howled, whitening the entire landscape to a state of folded cloth. As John Aitchison, our cameraman, and I arrived at the glacial head of a valley, the snowfall grew so thick that as it blew in waves across us, entire hills vanished entirely, Tipp-Exed from existence for minutes at a time. It was, therefore, quite some time before we realised that a harem of wild horses were feeding just a few metres in front of us.

The horse is one of the world's most familiar animals. 'Horsy' appears to be reliably one of the first animal words learned by many children, and horses have made their way into our lives through more routes than any other animal on Earth. Racing horses. Cart horses. Farm horses. Domestic and pet horses. The horse is an institution, and especially so in Britain. And yet, there is one arena where the horse now appears alien to our native sensibilities – and that is as a truly wild and native animal.

The sight of the wild horses in Mongolia was unearthly and spellbinding. They vanished and appeared, repeatedly, like a flickering set of ghosts. In snowfall almost a metre deep, the lead stallion had punched below the drift and was ripping dense grasses from below. Two bedraggled foals (adorned by a pair of magpies, happily collecting their fur to line nests) hid behind their shaggy fringes, as the snow whipped their faces. Alongside wild asses, musk-oxen and saiga antelopes, wild horses can endure some of the harshest conditions of any herbivore, long after the deer or even boar have moved out.

Horses are masters of endurance. This means that before their widespread extirpation, horses would have been amongst the most adaptable of Europe's large grazing animals. Pliny the Elder recounts large harems of wild horses living above the snow line in the Alps. We know from the observation of horses in recent European rewilding projects that horses can thrive in high mountains, low wetlands, grasslands, saltmarsh, open wood pastures, larger forest clearings, and diverse shrub-lands. There is, however, one habitat that horses shun – and that is dense woodland. The most obvious explanation for this, although hard to test in the present day, would be to limit the risk of foals being ambushed by wolves.

The fate of this ultra-adaptable animal as a truly wild species is, in one way, straightforward. Over many millennia (humans hunted horses, with specialist spears, as early as 400,000 years ago), we wiped out populations of wild horses across most of the world. And yet, at the same time, the wild horse, in many ways, lived on. Traditionally, the wild horse has been regarded as a single species, divided into just two distinct subspecies. In Britain and much of Europe, the tarpan (*Equus ferus ferus*), famously depicted in cave paintings in Lascaux, has been historically classified as a single subspecies. In historical times the Przewalski's horse (*Equus ferus przewalskii*), known now to the Mongolians as the *takhi*, or 'spirit horse', was believed to roam across the Mongolian and Russian steppes. Debate continues as to whether this more easterly of the two horses is, in fact, a different species from that which occurred in Britain and much of western Europe. Yet the concept of merely two types of wild horse may hide a far more complex truth.

Horses, like zebras, are perhaps more likely to have haunted a large range of habitat types; each varying in their design, or ecotype, based upon factors such as climate, vegetation, altitude and seasonality. Indeed, recent analysis of mitochondrial DNA in established breeds of domesticated horses reveals that across Eurasia, at least 77 different mare bloodlines are responsible for the diversity of horses we see today. As domestic herds spread across Europe, they, in turn, are most likely to have attracted the attention of remaining wild stallions, or mares, and thus the genetic diversity of original wild horse clusters would have been preserved. Therefore, quite how many original species, or subspecies, of wild horse once existed, has not yet been definitively answered – and it may never be.

What has happened, however, is that much of the horse's 'wildness' has been handed down, in parts; each encrypted and spread across the European landscape. From the hardy Yakutian, a native of Siberia that has probably changed little since it shared the steppe with mammoths, to the wetland-savvy Konik, today's ancient-breed horses and ponies preserve as much wildness as tameness.

Today, the word 'pony' conjures images of cuteness, domesticity and being ridden – yet the horses were never consulted on the word's application. Genealogically, in Britain, the Exmoor pony, whilst far from unchanged through millennia of domestication, is still considered the most similar in structure and ecology when compared to ancestral wild horses. Far hardier than most breeds of cattle, let alone sheep or deer, Exmoors can survive blizzards, repel driving rain with their heavy eyebrows, and eke out a living from the toughest of grasses, thistles, holly and gorse in the very dead of winter. Their lineage remains something of a mystery in terms of their genetic connection to the ancestral tarpan-type species of western Europe, but their morphology – or physical characteristics – are strikingly similar.

The famous cave paintings of Lascaux, in south-west France, etched by our hunting ancestors, depict, with great clarity and attentive observation of detail, the wild animals that lived around at the time. Here, drawings of wild horses, with dense fringes cast over their eyes, stocky, thickset bodies and scruffy black manes, are strikingly similar to animals like the Exmoor that persist until this day. And thus, unlike the now forever-vanished elephants and rhinos of Britain, much of the wild horse survives. It might be fair to say that the wild horses were not rendered entirely extinct, but rather 'eroded' over time, wild populations being wiped out as more and more domestic populations came in. Genetically diluted and changed the horse may forever be, but in its shape, character and sociability, the wild horse lives on.

The nativeness of a species to Britain, too, is generally defined around its presence on our island since the last Ice Age – and since we became an island in the first place. Pre-domestication bones of wild horses, however, have been found in Neolithic tombs, dating as recently as 3500 BC. This, at least, renders the horse a lost native – and a native whose many wild skills have, in an age of ponies, racehorses and petting, become largely forgotten.

In their wild capacity, horses were perhaps amongst the most adaptable of all ecosystem engineers – and this would have rendered their effect – in their original numbers, prior to their decimation, or domestication, by ourselves – amongst the greatest and least credited in the animal kingdom. Beavers are tied to water and trees, our ancient aurochs showed a marked attraction to lowlands and floodplains, and even wolves and bears to suitably rocky denning areas – but there are very few landscapes, except dense woodland, which cannot be lived in and shaped by free-roaming horses.

What's more, horses are hard to predate. Most prey of the grey wolf, Eurasia's dominant hunter of large animals in recent times, have at least one obvious weakness. Horned animals, such as cattle, are heavy, dangerous at the front, but not enormously agile, or powerful at the rear. Male boar can gore an opponent, but females and young have few defences. Wild horses, however, are extremely well adapted for predator defence. They have speed, muscularity, a devastating kick, a ferocious bite and, above all, a very tight-knit social structure. Generally, their main weakness lies in the fragility of their foals. And this brings us back to the wild horses of Mongolia.

When first reintroduced into the wild after a breeding programme in the 1980s, the Przewalski's horse proved, at first, extremely vulnerable to predation. Unfamiliar with wolves, many foals and younger animals fell prey to Mongolia's steppe-hardened packs. But, over time, this began to change. As the cohesion of harems was rebuilt – each dominated by a stallion and breeding female – the behaviour of the horses began to radically alter. Stallions quickly identified wolves as predators, and both stallions and mares have, in Hustai, been observed not only to close ranks around newborn foals, but also to attack, kick and even trample wolves. The sheer power of an adult horse's hind legs presents a formidable challenge for wolves, which, for all their social guile, are not invulnerable. In their first week of life, wild foals in Mongolia remain vulnerable to wolves, but after this time the risks reduce considerably. At present, in Hustai National Park, it is the large herds of red deer, not horses, that fall prey to wolves instead. The indomitable nature of the horse as an ecosystem engineer, therefore, might be seen as a combination of its adaptability and power – the very two attributes that have earned horses our admiration for millennia.

Wild horses form harems, which are often mutually exclusive of bachelor herds. This means that unlike other social ungulates (wildebeest, for example), prime grazing estate is often dominated by only a limited number of animals, in one place, at one time. Each spring, the harem's stallion will defend his place in the group against intruding males, seeking the right to mate with not one, but all of the females in the group. These fights can be sudden,

chaotic and fierce in the extreme. Males that are well
matched will often sidle up to one another, blowing air
through their nostrils, heads lowered; their entire heads
and manes thrumming with simmering energy as they size
each other up. At first, a few swift back-kicks, dealing
nothing more than enormous bruises, will generally decide
who is in charge. But failing this, conflict between rival
stallions can rapidly escalate. Ever fiercer back-kicks give
way to full-fronted, two-legged sparring matches, as both
rear on their hind legs. Eventually, if still no winner
presents, teeth clash and lock, often sinking deep, piercing
mane and neck flesh in a sometimes deadly throttle. These
fights determine everything in wild horse society; deciding
the future of the entire group, and its progeny. Once a
stallion is established as the leader of a harem, and has
mated, younger males, approaching adolescence, will often
be chased from the group. They may then form bachelor
groups; relegated to the less nutritious parts of a valley,
grassland or wetland. These mechanisms are fascinating in
and of themselves, but they also act to spread horses out
across a landscape, rather than focus enormous 'herds' in
one place at one time. So whilst domestic horses in fields
can, therefore, 'ruin' a habitat, free-roaming wild horses
have their own mechanisms for not denuding a landscape
beyond ecological repair. Instead, they serve a range of
crucial ecosystem functions.

When grazing in grasslands, horses are high-fibre
herbivores. They tackle and weed out coarse grasses,
thistles, thorn bushes and scrubland, leaving behind, in
many cases, finer forage for other herbivores such as deer.
Horses take out a lot of the rough grasses in a landscape
during the winter months, thus freeing up new plant
species to colonise come the spring, and thus, in moving
on from a grazing area, diversify the variety of flowering
plants. In Hungary's Aggtelek National Park, where Hucul
horses move through the landscape and then on into new

valleys and pastures, the lawn left in their wake transforms, weeks later, into a floral mosaic rarely seen in Britain. Indeed, its butterfly meadows far eclipse any of our own; home to numerous species assigned separate habitats here in the UK due, in part, to the rich variety of foodplants and nectar. Here, the habitat of species as varied as large coppers (feeding on water dock and meadow in the wetter parts of the valley) and purple emperors (their caterpillars feasting on sallow, their adults seeking nutrients from horse dung) exist within the same landscape. Over time, enormous dung piles, as they break down and change the nutrient composition of the soil, give way to verdant clumps of nettles; the foodplants, in turn, for yet more caterpillar species.

Of course, with true 'wildness' departed, there is always another ecosystem engineer present in the preservation of such meadows: ourselves. But by using 30 horses, grazed across one kilometre at a time, the Aggtelek's human conservationists have proven how powerful horses would once have been in shaping our grasslands as they moved endlessly throughout the landscape, following the changing seasons and the changing abundance of forage, and avoiding, where possible, the attentions of wolves.

There is even more significance to dead grass removal, however, than merely freeing up a wider range of finer grasses and plants to take root. Dead grasses can, over time, become a significant fire hazard. British conservation, with surprising consistency, advocates burning to remove dead matter. But dead matter only burns if it has attained sufficient density; a density thinned, or removed, by the presence of wild horses.

Where concentrated in preferred feeding areas, wild horses, as we found in Hustai, are able to create a number of microhabitats side by side. One of these is the 'horse-lawn'. Areas of dung-fertilised prime feeding ground, often beside a river or stream, these tightly cropped feeding areas drew a range of other species, not least dung beetles. Red-billed choughs and wheatears followed the harems, able to find not only an abundance of ground-dwelling beetle fauna, but also the cropped conditions that allowed that fauna to be readily detectable. This 'lawn', however, whilst cropped, is not the single-species 'green' that we see in so many of our lifeless farmland fields; all nuance, and most grassland flora, removed through penned grazing. It is, indeed, a type of 'shorn' meadow, which, on closer inspection, is still very much teeming with small flowers, fine grasses and invertebrates.

In Britain, we can still appreciate the merits of the horse-lawn in that one area where these animals are still left to shape large areas of largely original habitat: the New Forest. Here, around the edges of predominantly deciduous pasture woods, you can often be startled by the incongruous sound of 'open country' birds, such as lapwings or curlews, around the very edge of, or within, the forest itself. Yet these species are not using crops, fields or forestry clearings; they are feeding, and nesting, within horse-lawns.

Here, the cropped sward allows woodlarks and lapwings to readily access their prey. Wet streams, flooding areas of the horse-grazed pasture edge, also create the perfect conditions for lapwing chicks to feed. The woodlark, indeed, is perhaps an even more perfect example of an ancestral 'horse bird'. It sings in scattered, often dead trees close to the woodland edge, feeds in short glades and nests in low, tussocky grasslands. Small nuances such as bracken clumps or even large tufts of heather provide the female with enough cover to hide her nest on the ground. The actions of horses here allow species like the woodlark a hinterland habitat between trees, coarse and open ground.

Across Britain and Europe, you can still get atavistic glimpses of other 'horse birds', too. The yellow wagtails nest in dense, tall grassland, feed in cropped, damp grassland and often follow horses to find disturbed insects on the ground.

Corvids, such as the jackdaw, magpie, carrion crow and raven, are all intrinsically evolved to seek coarse fur to line their nests come early spring. Having watched Mongolian magpies pluck significant quantities of hair from hapless wild foals, in order to line their nests, it becomes apparent that the ancestral 'lining' of British nests would not have been imported sheep, as it is today – but native horse.

There is however one bird, beloved and familiar, whose dependence upon horses, in particular, is even now profound. In early 2021, when appealing across the UK, via social media, for large densities of nesting swallows, it was striking how many of the largest clusters, and most persistent dense populations left in Britain, were strongly associated with horse barns. Today, swallows gravitate to horses in confinement – once, they would have done so in the wild. And just as swallows still follow organic cattle herds in our healthier meadows and grasslands, so, in the future, we might see them once again hawking over horses – a familiar animal whose dung, and the insects it attracts, can dramatically aid the survival of one of our most beloved summer visitors, at the end of its 11,000-kilometre odyssey to our shores.

Scrub-grassland is amongst the richest of all habitats, creating a complex hinterland between patches of dense thorn and an array of grassland types, from meadowlands to pasture, which provide feeding and nesting habitat, both for those bird species that hide in thorn, and for others that spend their entire lives nesting, hiding and feeding in grassland. Horses achieve this by tackling not only the toughest of grass matter at their feet, but also the toughest of bushes and trees. Anyone who has marvelled at a New Forest pony tearing strips off a holly tree in the dead of winter, or gulping down luscious quantities of bramble or gorse, will appreciate the role that horses play in battling back some of the most formidable of our native bushes and trees. In doing so, horses act to create a mosaic habitat; simultaneously freeing up flowers to grow amongst grasses, and protecting grassland glades from complete succession into dense blankets of thorn scrubland or unbroken canopy woodland. The horse is a prime architect of balance in a landscape.

Today, scrub-grassland is amongst the commonest of 'combination' habitats maintained by human hands, often at considerable cost, on nature reserves that, being so small, must replicate natural processes in small areas. Humans, however, rarely do such a good job. In many of our small nature reserves, the complex 'scrub' elements, ranging from aspiring birch saplings to vital clusters of dense bramble – the home of many nesting passerines and hedgehogs and a vital nectar source for butterflies – are weeded out as undesirable. Pure grassland is often given preference – and this leads, over time, to an ever more specialised, manicured fauna. Over time, the natural jumble of habitats best suited to the broadest spectrum of wild colonists, is lost. Horses, by contrast, do not walk through habitats with fixed management strategies in mind. They tackle vegetation in the half-deliberate, half-random way you might expect from a species with a mobile, harem-led structure where feeding is often interrupted by fighting, mating, playing and drinking. The result is a habitat structure far more ramshackle, diverse and alive than many of the more orderly landscapes we have in Britain today. These 'soft-edged' habitats, vanished from many of our lowlands, and even more of our uplands, pack a range of woodland, scrubland and grassland into one place. And as we look to restore our lands in the decades to come, it is to such myriad chaos that we might look for inspiration.

Those planting woodlands from scratch, in Britain, restoring from zero the woodlands we have lost, face something of a quandary with horses. Horses can, and do, eat trees – they can kill saplings, and even, over time, adult trees. In a replete ecosystem, this is an essential role; by leaving standing deadwood within a scrubland ecosystem, horses radically increase the overall biodiversity of a habitat. Yet, if regrowing a habitat from scratch, it may be some time before the beneficial role of horses in an ecosystem can, once again, be realised. In many of the fenced rewilding projects seen in Britain thus far, such as the Carrifran Wildwood (a fenced glen in the Southern Uplands), it has taken 20 years for woodland to grow, like a phoenix, from the barren, overgrazed land. Now a glorious jumble of birch, alder, oak and treeline juniper, the glen is once again alive. In time, however, it will become robust enough for the custodians who once roamed here, and once again the wild horses that once helped shape our hillsides and coasts, our rivers and grasslands may integrate once again into their ancestral home.

If wild horses are the missing, yet essential, component of our grasslands, from our coasts and hills to river valleys and scrub or woodland edges, there was once another animal, even more mighty and powerful, that shaped the majority of the British landscape that was once heavily wooded. It was an animal whose size we can find only in the Indian bison, or gaur, possessed of a muscular majesty, and it is at once utterly alien and familiar to British sensibilities: the aurochs.

Of the cornerstone species long forgotten to our shores, within the Holocene, and since the last glacial period, none was more impressive than the aurochs. The wild ancestor of domestic cattle, this two-metre-tall bundle of muscle would have challenged even a large pack of wolves. Some have even theorised aurochs were so dangerous (like the

present-day gaur) that predators such as wolves may simply
have avoided them entirely; choosing instead the plenitude
of red and roe deer, European elk and other mid-sized prey
– species that could be taken with far less threat to life or
limb. The fate that befell the aurochs was no different to
that of the wild horse. Over time, wild populations were
hunted to extinction. As with much of Britain's fauna, this
would happen far earlier on our post-glacial island than in
Europe. In Europe, aurochs persisted until 1627, when the
last died out in the Jaktorów Forest of Poland. In Britain, it
is believed they vanished as long ago as 2,500 years before
the present.

The aurochs was a giant animal, with giant ecosystem
impact. We can trace that impact in many ways. Indeed, it
has left far better traces than many other vanished animals.
Its habitat, for example, has been surprisingly well examined.
In a comprehensive analysis of fossil records, biologist
Stephen J. Hall tested the long-standing hypothesis that
aurochs were, predominantly, the governing megafauna of
our lowland floodplains and woodlands. He found that the
vast majority of fossils of aurochs (in contrast to cave- and
crag-favouring species, such as the wolf or brown bear,
which often favoured upland areas) came from our lowlands;
in many cases, contemporary agricultural areas, long-cleared,
where woodlands, or floodplains, or both, would have lain
in the past.

Most animals, however, towards the end of their tenure
on our planet, are often driven into some form of refugia.
Whether the aurochs was always a floodplain species, or
whether, over time, it simply became harder to hunt them in
floodplain forest, and thus they persisted here for longer,
seems difficult to determine. Most of the world's few
remaining wild cattle – such as the Indian gaur – are
woodland animals, but can inhabit and thrive in both
wetland and drier woodland ecotypes.

Given that 20 per cent of Britain was once wetland, marsh or floodplain, and vast areas, such as the Humber, East Anglian Fens, Cheshire mosslands and Somerset Levels, were shaped entirely by the actions of fresh water over the seasons, the role of aurochs in our wetlands would have been profound.

It has often been observed that floodplains require grazing in order to maintain their maximum biodiversity and the openness beneficial to many of the species that inhabit them. Left alone, bereft of large grazing animals, areas of open wetland, reed-bed, sedge fen and other waterlogged habitats can be quickly colonised by trees such as alder. This process of regeneration is, of course, entirely natural – but it is also entirely natural that such growth should be contested.

Aurochs, armed with curved horns, were grazing animals, and their diet, in spite of their preference for wetland and wooded areas, is believed to have consisted predominantly of grass matter, supplemented by the leaves of woodland trees and scrub. This would have allowed a dual structure to form in our lowlands, with grassland glades kept partially open and free from scrub, whilst trees and shrubs would have been pruned back fiercely but not, necessarily, broken or destroyed. The formidable horns possessed by the aurochs, especially the black males, give us robust hints as to their purpose – indeed, observations of old-breed horned cattle today tell us the same story: the horns were used, when feeding, like a scoop, to pull down branches and rip off twigs and leaves. This would have led to a powerful coppicing effect in our woodlands, long before foresters invented similar, more precise methods millennia later.

In areas such as the original river valleys of the Norfolk Broads, where original floodplain vegetation has been left unchecked by any floodplain 'gardener', vast, unbroken alder woodlands have formed. These, in many cases, are surprisingly species-poor, being, in truth, only a part of a wetland ecosystem once gardened by beavers, rotavated by wetland boar, debarked by large herds of red deer – and seriously disrupted by wetland cattle. Sometimes, it is only through the absence of long-lost stewards that you can detect what happens once they've gone. In contrast, in Poland's Biebrza Marshes, where marsh cattle are grazed across the river valley over the course of the year, moving, or being moved, with the seasons, we get a far better glimpse of the diversity of vegetation structure – and biodiversity – once effected by free-roaming cattle in our marshlands.

Given that we cannot travel back in time to study the full range of ecosystem actions carried out by the extinct aurochs, a great deal of reconstruction of how free-roaming cattle, wandering in small family groups, would have impacted and shaped our landscape, has come from far more recent rewilding projects, like those seen on the Knepp Estate, in Sussex, where old-breed horned cattle have been returned to roam the land. At Knepp, Old English longhorns have been the 'aurochs' of choice, as documented in Isabella Tree's international bestseller, *Wilding*. Given once again the run of the land, and a wide array of habitat choices, and being freed from their farmyard heritage, the longhorns have revealed much about how wild cattle would have, and could again, shape the British landscape.

It is fascinating that simply by moving, eating and defecating, cattle justify their role as cornerstone ecosystem engineers. Compared to roe deer, which can only effectively transport around 28 types of plant seeds around the landscape, cattle can successfully vector more than 230 species of plant, carried and re-seeded from their fur, hooves and gut, often complete with a healthy dose of natural fertiliser. This, in part, reflects the manner in which cattle feed. Lacking an upper set of front teeth, they are relatively coarse feeders, wrapping long tongues around grasses and flowers, unable to be as selective as a smaller, browsing animal possessed of finer dentition. This in turn results in a wide variety of grasses and flowers being ingested – and, days later, distributed as widely across the land as that animal has roamed. This clumsy yet elegant seed dispersal mechanism would once have ensured that plants and flowers were vectored across our wetlands and lowlands, where, no doubt, they would also have taken root in beaver dams and boar wallows, as well as the damp, expectant soil. In addition, even the footprints of cattle can develop into miniature ecosystems in their own right; micro-ponds for newts or, as can be seen in the Outer Hebrides, shallow earthy depressions used by nesting waders such as lapwings.

Like most wild species, it is likely that aurochs had their favourite trees, trees they would have tolerated, and other species they would have avoided entirely – and this in turn would have led to greater variety in the lowland landscape, and the structure of its woodlands. At Knepp, for example, the longhorns' preference for damp-loving sallows, which would have been common in our floodplains or any lowlands with damp soil, gives clues as to how certain trees were selected. The leaves of sallow contain salicylic acid, a natural anti-inflammatory that is thought to relieve worm burden in cattle – and would have done so long before soil-toxic medications were invented to solve the same problem in our domesticated breeds. Even apparently unfussy eaters are rarely random in their choices. Likewise, it has also been observed that after giving birth, Knepp's longhorns chew on nettles rich in iron. This reminds us how quickly, and selectively, our onetime wild cattle might have switched diet – and, in doing so, carried out numerous different acts of pruning, gardening and grazing, with a short space of time, and within a small area of the landscape. This is likely, again, to have resulted in the 'soft-edged' habitats we find more commonly in eastern Europe, where cattle roam at the ten-kilometre level in some areas, exercising numerous different choices as they go; choices that incrementally shape the landscape and increase its heterogeneity.

How wild cattle and horses once interacted, overlapped and the exact range of habitats they once used, cannot be deduced from the fossil record alone, especially as horses were, it seems, removed earlier from the landscape, in most places, than aurochs were. However, from the available fossil record, and the preferences of our old-breed horned cattle (when given a choice), we know that cattle were woodland animals, whereas horses worldwide are observed to avoid dense woodland, preferring open or semi-open habitats. In some areas, however, it seems very likely cattle and horses would have overlapped: in our rich scrublands, tree-studded grasslands and at the woodland edge, and, perhaps most of all, in our wetlands. Just as the Okavango in Botswana is, to this day, the ultimate nirvana for large herbivores, from zebra to elephant, because it provides the richest feeding environment of all, so Britain's onetime vast floodplains, from Avalon to the Fens and the Humber, might once have been blackened by aurochs and silvered by horses, all feeding, in complimentary fashion, side by side. And what a sight that would have made.

Today, the rich possibilities of treating cattle and horses as wild animals, of adopting the best of old breeds to create a new future for wild grazers, are just beginning to be realised. Projects like the wildland at Knepp remind us how quickly the competition of small, breeding groups of cattle and horses against vegetation succession can create a rich and varied paradise, now home, in just 20 years, to the greatest density of breeding songbirds in Britain. But all too often, we lack imagination in how we might use these familiar animals to the good of the environment, and the restoration of our lands.

Whereas the grey whale has been lost to all living memory in the UK, cattle and horses are almost too familiar: we forget that they, too, are supposed to be wild, or mistake putting smaller numbers in our fields for effective conservation. If we are to restore the land, cattle and horses must roam free. When they do, these onetime farmyard creatures may reveal, once more, a wildness innate in themselves and in the landscape all around.

CHAPTER SEVEN

Trees

It is said we must plant more trees. But which ones? And should we plant them at all – or let them plant themselves? Whilst it has become the go-to advice of those saving the planet that we need more trees, there are in fact many ways to misunderstand trees; to enforce them upon landscapes where they do not belong, to cluster them in formations that nature does not understand or put them where they do not want to be. We have grown to favour certain trees, to develop favourites, to develop a sense of purity of what grows where. So how do wild trees behave? How are trees supposed to grow, to thrive – and to shape our world?

Trees are wild beings that grow ever more unpredictable, haphazard and diverse with age. They attain more variety than any other group of flora or fauna in Britain. Some, like alder, ash, or oak, excel in death – bringing life to whole new orders from beyond the arboreal grave. Trees were not once orderly, planted affairs but chaotic giants – and if we are to rewild the world, then it is to this state that our trees must return.

The ecosystem function of trees themselves is well known, and too well expressed elsewhere (such as the wonderful *Hidden Life of Trees*, by Peter Wohlleben) to be the subject of this chapter. Instead, as we decide how to reforest our country, and look to a more wooded future for our island, it is worth remembering how wild woodland looks, how it shapes the landscape, and how our flora and fauna are intrinsically adapted to move between different native tree species, at different times of their lives, to shape trees, live in them and within their decaying limbs, across the generations.

Wild woodland plants itself – or is planted, in an apparently haphazard fashion, by a series of furry and feathered foresters. The jay, amongst the most important of all tree planters, becomes the first character in our wild tree planting story, as it is the prime architect of oak woodland – and of how oak woodland forms over time. And as oak is the most biodiverse of all British trees, the role of this one colourful bird is hard to overestimate. Jays collect healthy, ripe acorns that oak trees drop as their fruit come the autumn. Whilst some jays may cache a handful of these close to the oak in question, others, in carrying larger numbers in their crop, disperse acorns far further afield. Amplified, these actions begin the creation of a chaotic woodland landscape. A group of 65 jays within a woodland, for example, can, over the course of four weeks, disperse as many as half a million acorns across a landscape. Many of these will be 'planted' many miles from the source oak. Even 200m or so from the raided oak, up to 5,000 acorns can fall per hectare. And so begins a new generation of wild woodland.

It has been observed by many foresters and ecologists, notably the Dutch ecologist Frans Vera, that oaks, from dropped acorns, do not prosper in shade, where they can often be eclipsed in the early stages of their growth by species such as hornbeam, or beech, that are shade-tolerant. In addition, jays often choose loose, disrupted soils in which to bury their acorns, with each jay remembering only those acorns that it has itself buried. Often, these burial sites will be in scrubland, or grassland – and the soil disturbance may, historically, have been caused by other keystone species in our landscape; large grazers, such as horses or cattle, or boar. Both in terms of its intolerance of shade, and how and where it is planted to begin with, by jays, the oak – one of the most powerful cornerstone species of all – is born and thrives best as a creature of the light.

A sapling oak, growing openly in sunlight, however, is a delicate, and edible, thing. Large herbivores, including deer, can destroy many in their earliest stages – and oak, therefore, grows best when protected; bursting through a fortress of thorns: of gorse, blackthorn, hawthorn, bramble or wild rose. The ancient saying from the New Forest that 'the thorn is the mother of the oak,' can be seen in those areas of our landscape where jays, scrubland and oaks still thrive as a symbiotic force. On the rewilded Knepp Estate, in Sussex, the vast majority of new oaks grow through thick thorn scrub, which acts to protect the sapling against being browsed to death early on in its 500-year career as a tree.

The means by which the oak most readily colonises our landscape is a fascinating reminder of how divorced many of us, even many foresters, have become from our most familiar trees. Instead of being planted in serried ranks, the natural biological inclination of the English oak is to fall from the sky, as a fruit, dropped by a bird, into a dense protective shell of scrub. It is this combination of randomness with design that, over time, leads oaks to stud landscapes with their gnarled presence, growing far enough apart that they do not outcompete one another. Five hundred years on, the New Forest's oldest wood pastures, like Bramshaw, reveal to us how such woodland eventually looks; an oak-studded pasture, filled with light and life.

When we think about wild woodland, and its return to cover ever larger areas of Britain, as it once did, bringing life back to the land, it is important to remember how that wild woodland once formed. Oak-led wood pasture, which may once have covered vast tracts of lowland Britain, is unlikely to have begun life as a dense growth of trees, but rather as a tree-studded landscape growing through thorn; resistant to the pressures of herbivory. Over time, as those sapling oaks grew ever larger, they would have long outlived the brambles and thorn trees from which they once grew. Over time, in some areas, such oaks, if planted profusely, would have grown together into clusters, or woods – in others, where planted more sparsely, or if grazing pressure was higher in that landscape, to dot our grasslands like giant monoliths, more akin to the large acacias of the Serengeti. By contrast, a dense, continuous forest of oak is rarely a wild phenomenon, but more often, the expression of a wood planted by those newer foresters, ourselves, to grow the timber for ships or buildings, in one place, using one species of tree – in far more recent times.

If the kind of woodpecker-drummed, beetle-riven and fungi-encrusted oak-wood pastures we can still find in areas like the New Forest are the *end result* of wild woodland planting itself, there are many magical stages along the way – many of which have been sidelined, or forgotten, by conservationists or foresters today. One of the most important of these, within any thriving and self-governed landscape, is the Great British Shrubland. Now relegated too often to outgrown, forgotten hedges, airfields or the edges of power stations and industrial estates, shrubland, or 'scrub', would once have covered large areas of our landscape. Kept low and semi-open by the grazing pressure of large herbivores, deer and the rootling of boar, its diversity would have been extraordinary and its sound deafening. Today, the densest chorus of birdsong in England comes from the thorn scrublands of the Knepp Estate – but scrubland was once a true habitat, not merely a refugee relic of our countryside.

In keeping with the fact that the mighty English oak thrives best in sunlight, many trees that we now associate with hedges (hawthorn, blackthorn, buckthorn), with coppice (hazel) or with wild fruit picking (elder, crab-apple, and wild rose), are all intrinsically the key components of a rich scrubland landscape. Whilst we may have brought hazel into our woodlands and coppiced it, hazel thrives best as a sun-fuelled scrubland tree. Whilst we may have become accustomed to finding elder or crab-apple in hedges, most of these existing only since the 1700s, both are scrubland specialists; crab-apples, in particular, growing in wild groves in the New Forest, and other areas where horses are still present. The horses vector their seeds across the landscape. With wild boar once common throughout Britain, they, too, would have played an important role in transporting apples across the landscape, excreting their seeds intact and with a healthy dose of wild fertiliser to boot. Our thorn trees and fruiting trees once constituted a complex mosaic of scrubland – one to which many of our native fauna are intrinsically adapted. From the hazel dormouse, which moves through networks of fine branches to find wasp galls or fruits, to the long-tailed tit, which often feeds and forages for nesting material within blackthorn but nests most often in dense bramble, many of our species are perfectly adapted to a world of sunlight-fuelled thorn and fruit.

When it comes to the appreciation of wild woodland, and allowing it to recolonise our landscape, we must therefore appreciate the truly chaotic nature of thorn-lands, and their enormous importance to our native species. Indeed, a whole host of species ascribed to the farm – to the arable field or hedgerow – such as the vanishing turtle dove, or others, ascribed to woodland, such as the nightingale, are originally the denizens of complex, interrupted and diverse thorn-lands, where glades meet dense fortresses of spiky shade. The turtle dove, for example, nests within dense, tall thorn bushes, but forages in disturbed, sunlit soils for the seeds of fumitory, chickweed, knot-grass and fescue. The boar, and the hawthorn, are perhaps its most true wild friends – no longer the farm, where dense hedgerows have often been eradicated, and the arable weeds that fed it for centuries now clinically removed from the landscape, along with 97 per cent of British turtle doves.

A host of other species, from a variety of orders, remind us that thorn-land was once amongst the most dominant and important of our wild woodlands. The black hairstreak butterfly specialises in towering castles of sunlight, ancient blackthorn – but disappears in shady woods, and can only be found along rides where blackthorns attain their full fortress-like forms. The blackthorn and hawthorn, if vigorously pruned in a hedgerow or garden, only respond by growing more virulently over the coming months. This ancient response to herbivory has far-reaching consequences for our native wildlife. The shy lesser whitethroat, for example, arriving into our farmland each April from Africa, inevitably seeks out the densest of flowering blackthorns in which to sing and nest, and its arrival each year appears to synchronise perfectly with the height of the blackthorn's blossom. Whilst our scrubland trees have been reformed, or rather, confined and straitjacketed to modern landscape features like the hedgerow, we are confronted everywhere with reminders that the British scrubland was once widespread, diverse and, whilst apparently chaotic, enjoyed a harmonious synchronicity between its maze of low, bushy trees and glades – and the animals within.

Whilst English oak, home to more known invertebrate species than any other native British tree, may be considered the greatest of our cornerstone trees, oaks are distinctly average trees until around 300 years of age. Whilst owls, woodpeckers, many beetles and fungi and a host of smaller moths and butterflies, such as the purple hairstreak, all thrive best around mature oak, there are also whole armies of creatures for whom the more humble blackthorn and hawthorn remain just as important. The nightingale, for example, thrives best in younger scrublands where aspiring hawthorns are still cloaked in bramble, nettle and wild rose. The red-backed shrike, formerly common in our countryside, makes hawthorn its larder for skewering beetles, grasshoppers, birds and lizards, as well as its watch-point, and often, its nest site. Cuckoos find many of their favoured caterpillars in thorn trees within grassland come May. After oak and willow, hawthorn and blackthorn are also home to more invertebrates than any other native; their presence in a landscape as important, in their first 50 years of life, as oak is in its last few centuries.

The importance of our fruit trees for wild animals is also easy to forget, as so many fruit trees have now vanished from our countryside that few talk of replanting them with the same zeal as creating new forests of oak, hornbeam, beech or Scots pine. Yet fruit trees lie at the very heart of any wild woodland – and co-evolved with our bears, horses, boar and birds for millions of years. The crab-apple is perhaps the best known of our wild fruit trees, but the dog-rose, elder, wild pear, rowan and especially bird cherry – the latter now rare in many of our woodlands – would all have been important components of our scrublands and wood pastures. Indeed, one of the striking phenomena about Britain's native trees is how, as a combined force, they provide virtually no hunger gap for our resident fauna as summer fades. From blackberries, rowan berries and crab apples in August to haws and hips in September, all the way through to bullace (wild plum) and ripe hazelnuts in October, and to ivy berries in December, many of our resident species have a continually evolving seasonal menu from which to choose. However, that menu only works in diverse landscapes where all of these light-loving trees coexist. As we look to rebuilding our woodlands in the future, it might therefore be wise to pay as much as attention to fruit, and to nuts, as a bear might; and not only to trees as carbon-sequesting units but in terms of their yield for the natural world.

The wild woodland of Britain would, in at least 20 per cent of our once undrained land area, have been shaped by another mighty force: water. Riverine woodlands, and floodplains, favour the growth of different trees – and of different communities within them. These were lands of varied willows and mighty black poplar; of thriving elder scrub, osier and damp-loving alder. Where water and the dams of beavers once dominated, wild woodland would have taken on a different aspect. Riparian woodlands have a character of their own, but again, their wildness is chaotic and self-governing – one cannot simply plant a wild floodplain woodland, unless one is a beaver, of course.

With so much of Britain drained, it is the river valleys of eastern Europe that give us a better idea of how woodland once worked in landscapes entirely submerged in winter by powerful rivers; landscapes that would then have dried, slowly, as the summer regained control. At the height of a wild river's flood, few trees can withstand its flow. Crack willow and white willow, black poplar and alder are amongst the few trees that actively thrive under such conditions. Others, like damp-loving elder, can also do well. To see oaks thriving within a wild floodplain, however, is unusual; beech and ash need even not apply. In many ways, European floodplain woodlands are as special as rainforests – indeed, they are more specialised, too. Here, processes such as flooding interact with animals such as beavers. The trees are far from a passive force – but caught in the middle they can be. Willow and aspen, favoured by beavers, can be fractured, chewed and gnawed to within an inch of their lives – yet froth with new shoots months after the most severe act of beaver coppicing. Other trees within this flood-riven, beaver-nibbled domain, such as elder, are toxic – and left alone. Alder is often shunned by beavers and thus is the tree most able to form tall, dense and continuous woodlands on a floodplain. It is indeed interesting how many swamp-nesting species, such as the white-backed woodpecker found in old-growth woodlands in Europe, choose to place their nests in alder. It is, perhaps, a safer bet for a home, after all, than one of the beaver's favourite food-stuffs.

When large rivers like the Pripyat expand and contract each winter, the very shape of trees closest to the river's main course is transformed. Willows, for example, can sometimes grow to become giants but are often battered into bush-like forms, as much by the winter flood as by beavers. Black poplar, however – now almost forgotten in large areas of our countryside – stands tall; a monolith proclaiming that you are standing on a floodplain. This is amongst the richest of all trees for invertebrates; its craggy bark providing infinite, miniscule nesting apartments, and a dense canopy for singing and nesting birds. It is interesting, when we talk of planting trees, how many of these keystone giants remain forgotten – along with the processes that once created and maintained their majesty. We have also forgotten other animals, like the mighty Eurasian elk, which once hassled, coppiced and often nibbled away at the birches, aspens and willows around our rivers. These enormous four-legged secateurs would, in turn, have been kept mobile and hunted by wolves, but nonetheless exercised a powerful pruning effect upon these trees. It is perhaps little surprise that the willow – accustomed to being battered by rivers, coppiced by beavers, browsed by aurochs and nibbled by elk, to name but four titanic natural forces – is amongst the most resilient and unkillable of trees.

Floodplain woodland would once have covered large tracts of lowland Britain. Whilst beavers would have created glades and coppiced, and aurochs would have grazed and kept areas of floodplain in a state of low verdure, the expansion of rivers in winter would have driven aurochs or horses further away from the water's course, whilst meanders and oxbows would have created other areas where large tracts of alder and willow would have developed into great woodlands over time. Beavers, in undiminished numbers, would have transformed these woodlands from even carpets of trees into a rich mosaic of dams and ponds which, in drying over time, would have opened up new meadowlands within; and new forage for the aurochs. Today, when we do encounter floodplain woodlands, as can be found in the unbroken alder-woods of the Norfolk Broads, they are often peculiarly silent. This is because they are also peculiarly unnatural. Two ecosystem giants – beavers and cattle – which once opened them, have gone, leaving in place silent walls of alder without the disrupting forces that once made them resonate with life.

If we have forgotten how wild our thorn-lands, our fruit-groves, our river valley woodlands once were, there are better hints left on our battered island of the majesty of another kind of woodland – once which would have functioned under different rules again: the Atlantic rainforests of our western coast. It is not a revelation to anyone that the west coast, especially in Wales and Scotland, receives enormous amounts of rain each year – and has done, in all probability, through much of the Holocene. What is more easily forgotten, however, is that this weather system, unique to our island anywhere on the globe, is now watering a rainforest that is no longer there!

The western Atlantic rainforests of our island may have held a very different aspect, and wildness, to other parts of Britain. From the gnarled boughs of Exmoor's Horner Wood to the impressive remnants of ancient deciduous woodland that flanks the slopes of Meirionnydd, we can still find habitats where rainfall and trees conspire to create habitats so verdant, so moist, that, like the Pacific Rainforests of Canada, the dominant flora of the ground layer is greened with mosses, liverworts and lichens. Many of these ravine or gorge woodlands, growing on steep slopes, the moss-encrusted hazels bursting from green-carpeted rocks, remind us that in some areas of our country, climate, rainfall and steep inclines would all have favoured the complete dominance of deciduous woodland cover. These areas, neither, it would seem from the fossil record, the favoured haunts of aurochs or horses, nor suited to beavers, would have been true rainforest. Whilst many habitats have now vanished entirely from Britain in their natural form, it is likely that the woodlands we can still glimpse in the valleys of Dartmoor, Exmoor or the Mendips, and indeed as far north in western Britain as Crinan or Sunart, in Argyll – are natural expressions of wilderness. Even though their rocky outcrops would once have been wolf-loud, and their cavernous uprooted tree canyons the haunt of slumbering bears come the autumn, these rainforest fragments remain intact. Here, trees express their wildness in a very different way.

Celtic rainforest, as it is now colloquially known, is a mossy jumble of trees at their most verdant and intense; a place of specialists and speciation. Here, mosses, liverworts, fungi and lichens jostle – encrusting living trees as well as dead. Sessile oak, hornbeam and limes can prosper here, whilst a wide array of other species, such as ash, hazel and downy birch, all grow fast, fuelled by water and sunlight, and rot fast too, fuelling a world of decay specialists from fungi to beetles, woodpeckers and willow tits.

Britain's trees want to be wild in many different ways, and in each of the microclimates of our island, a new set of rules springs into being. None of these can be easily overwritten by planting, plastic tubing or forestry. Indeed, a true Atlantic rainforest cannot be planted or replanted – it is so ramshackle, chaotic and diverse as to be only able to grow by itself. If you wander deep into the remaining fragments of ancient rainforest in our island, it is to be surrounded by a fierce competition of trees – some, like rowan, bursting out of rock crevices where no human would conceive to ever plant a tree. These trees are engaged in the most fierce acts of competition, each pushing upwards towards the sunlight and only the most successful, like the towering limes of the Wye's steep slopes, eventually breaking into the canopy and in doing so, shading out others in their wake. This intensely competitive world relies upon chance. If a jay, or rodent, drops an acorn, this determines how likely it is to survive. Oaks can survive and even grow to dominate our rainforests, but many will be swallowed by the shade of small- or large-leaved limes or hornbeams. Like the emergent kapoks of the Amazonian rainforest, only a fortunate few will reach the top.

The specialism of some British tree species, such as the fascinating Llangollen whitebeam, reminds us that in additional to the generalist giants, such as English oak or white willow, which can be found across the British countryside even today, there were other trees that may, once upon a time, have been characteristic of specific rainforests now entirely lost to our memory. The Llangollen whitebeam is currently known from just one gorge in Denbighshire, and one quarry in adjacent Shropshire. The Avon whitebeam is known only from Bristol's Avon Gorge; the Ley's Whitebeam, the Brecon Beacons. A lot of these trees may once have been part of specialised rainforest communities, evolving in isolation from their peers in gorges just a hundred miles distant. It is such trees that remind us how peculiar, specialised and isolated from one another, some tracts of Celtic rainforest would have been. In addition, island communities of trees would each have held a different character. Arran, for example, is home to no fewer than three natives found nowhere else on Earth – the Arran service tree, and two species of the rose family, *Sorbus arranensis* and *Sorbus pseudomeinichii*. Isolated from rainforests on the mainland, these trees would have diverged from their peers over thousands of years. To find speciation on an island is not particularly surprising. To find speciation with individual gorges of Britain, however, reminds us rather than being one, continuous forest, our Atlantic woodlands would each have evolved as semi-separate entities, with different wild components in each one.

The wildness of trees would not only have changed in our rainforests, predominantly in the rain-fuelled Atlantic coast, but also, towards the north of Britain – as it does to this day. Whilst the great wood of Caledon – or, more accurately, the enormous woodland covering northern Scotland – was traditionally thought to be dominated by pine, it is now thought that this merely reflects those trees that survived into later times. Instead, it is likely that a rich carpet of birch, aspen, willow, pine, hazel, wych elm, rowan, oak and many other trees may have covered large areas of northern Britain.

In the growing number of glens and rewilded areas in Scotland where natural regeneration is taking place, birch is striking for its ability to recolonise the landscape. Many areas currently denuded by intensive grouse-hunting, and turned to blanket heather, or fell-side sheep grazing, rapidly revert to a sea of pale green birch once the processes preventing regeneration (such as heather burning, or sheep farming) are removed from the picture. Birch is one of the northern hemisphere's most remarkable and adaptable trees. Indeed, on travelling northwards in Scandinavia, passing from Finland and into Lapland – and the Arctic Circle – dwarf birches and willows continue to stud the landscape long after the last pine, spruce or fir woodlands have fallen south behind you. Were areas like Orkney, Shetland or the Hebrides ever to be recolonised by trees, all of the limited fenced experiments to date suggest that whether on windswept islands or the farthest northern island of Britain – Unst – low forms of birch are most readily able to colonise the landscape. It is interesting, therefore, that the woodland terms we ascribe to an area – whether 'closed canopy' woodland in southern England, or 'Caledonian pine forest' in Scotland – rarely reflect how wild trees actually grow, or what happens when trees are allowed to regenerate naturally, alongside natural processes. Generally, British woodlands behave in a more chaotic and varied fashion than the rules we set for them on paper.

Some wild woodlands, however, have become so rare as to be almost entirely forgotten. As so many of Britain's hillsides have become denuded right up to the native treeline, whether through sheep-grazing, heather burning or overbrowsing by deer, it can be near impossible to discover true montane woodlands in all their miniature richness; woodlands that are still a common sight in the uplands of south-west Norway. In montane or 'treeline' woodland, trees, given a chance, express themselves in entirely different ways. A broken sea of dwarf-shrubs and twisted, low boughs emerges – dominated by berry-rich juniper, rowan (more accurately known by its other name, mountain ash), dwarf birch, stunted high-altitude birches and low, montane willows. Here, the treeline fades out slowly, over hundreds of metres in vertical ascent, rather than stopping suddenly, as might a line of planted forestry.

So rare is this form of woodland in our country that until efforts to restore it began, only one recognised example remained – that of Creag Fiaclach, in the western Cairngorms. But the montane wooded world is one that we should not forget. In the uplands of south-west Norway, many species in sharp decline in our own uplands – such as the black grouse, merlin and ring ouzel – find a far safer and more varied refuge than the blanket heather or grassland they are forced to exist in here in Britain. Ouzels feed up on juniper berries after fledging, whilst the presence of birch buds, juniper and rowan berries on hillsides is vital for increasing the autumn and winter survival rates of black grouse. Merlins in Britain often nest on the ground in our uplands because we give them no choice – rendering them far more vulnerable to foxes. In southern Norway, merlins will often hunt rough grasslands for prey but nest safely off the ground, in old crow nests in low montane trees. The montane woodland provides sanctuary, variety and food – but only if we allow it to regrow. Other species, such as the bluethroat or Lapland bunting, all of which pass through

Britain each year and occasionally sing, or breed, in the Scottish Highlands, lack almost any suitable habitat to colonise within Scotland's vast montane landscape.

Rewilding our montane woodlands would transform the huge areas of lifeless, stony hills that may appear wild but are, simply, empty. From the Lake District to the peaks of western Scotland, montane woodland is the native habitat – however long it has been absent. Restoring it will allow trees to reveal to us formation and wonder that we never knew existed. And over time, these vital treeline woodlands could become repopulated with the animals that once roamed here; browsing their leaves, like elk, or grazing between them and on them, like wild horses. Yet, as with so many conversations we have about trees, and tree-planting, we forget that trees are communal creatures, for whom each community forms a rich yet distinct culture, over time.

Time, in fact, is the greatest asset to any woodland. From the stag beetle to the lesser spotted woodpecker, whole orders of fauna exist in Britain (many now critically endangered) for whom the action of time in trees is the only recipe for their continuance and preservation. If our trees are to be more than carbon-soaking units and form living communities, time is the last and greatest ingredient for any wild wood.

Indeed, there is more life in many dead trees than in living ones – and this strange state of affairs is what makes trees unique as organisms on our planet. Whereas the carcass of a large animal, brought down by wolves, may temporarily feed eagles or vultures, other carrion-feeding birds and even add vital nutrients to the soil, the ecosystem effect of animal decay is short-lived. By contrast, trees, as their limbs decay or fall and, eventually, their entire frames, can feed and sustain entire ecosystems not only for years but for centuries.

The process of deadwood decay is amongst the powerful drivers of ecosystem function – the final component of any healthy woodland.

A great number of our trees are only of moderate use to the natural world in their younger, living form. Ash, in particular, is noted for its relatively poor invertebrate communities, and is often not favoured by many nesting birds or smaller orders of moths, beetles or fungi. A decaying ash, however, is a different prospect entirely. Time rips open large cavities in ash suitable for owls or cavernous gaps at their bases where foxes or polecats can stash their young. Wood decay in its branches invite beetles, fungi and lichens to the party. Over time, the regular shape of living ash gives way to the more varied, accommodating and random shape that only a dying tree provides. And as ash dies, it continues to stand, often for many years – and in doing so, it continues to provide homes for more and more wood decay species, such as woodpeckers.

Open-grown trees, such as oak, decaying in wood pasture, can become over time invaluable invertebrate magnets in the wider countryside, which ten thousand newly planted, plastic-cased oaks could not replace in terms of ecosystem importance. Decaying, open-grown English oak is perhaps the single richest miniature ecosystem we have on the land. Fruit trees, such as crab-apple, cherry or plum, develop invertebrate communities of flies and moths that grow richer with every passing year. Yet is the collective impact of decaying and dead trees upon entire landscapes that is the most significant factor in an ecosystem.

In Poland's Białowieża Forest, a team of Polish ecologists, led by Andrzej Bobiec, studied the diverse effects of decaying tree communities and wood decay upon the landscape as a whole, in a study entitled "The Afterlife of a Tree". They found that within just one square kilometre of the oldest and least-managed parts of this primeval European forest, at least 100 trees died each year. The team

looked not only at what invertebrates moved into deadwood
and thrived, but also, at what stage of decay new invertebrates
arrived to the party. They found that on average, within a
year of a tree's death, beetles like longhorns begin to bore
into the timber. In the second to fourth year of decay, the
number of insects able to digest wood increases as the
wood grows softer and more palatable, whilst loose bark
begins to provide a home for larger invertebrates to hide.
Up to six years after the tree's apparent demise, longhorn
beetles and ants are now well and truly at home, whilst
ground beetles can now overwinter within the timber.
Between seven and nine years, as the sapwood lining the
tree rots to softness, ants, robber flies and many predatory
insects move in. And nine or more years from death, as the
heartwood rots, rove beetles, earwigs and earthworms all
move in for the feast. The more rotten a tree, the more
invertebrate life accrues to the banquet.

As trees grow ever greater in death, the number of
vertebrates that accrue to their boughs, some gnarled, some
fallen, also grows in number. In this regard, species such as
woodpeckers become miniature ecosystem engineers in
their own right; chiselling out desirable nesting cavities that,
on becoming their abandoned second homes, are colonised
by owls and a wide array of bats. Indeed, bats are one of few
orders of mammals who are, for the large part, dependent
upon dead or decaying trees for survival at all. Whilst in
recent times, some species of bat have moved into our own
apartments, it is woodpecker holes, hollow tree chimneys
and gaping cavities in ancient trees that provide. As decay
sets in, fallen trees continue to propagate new life. Root
plates become the favoured nesting haunts of species as
humble as the dunnock and as mighty as the eagle owl.
Honey fungus, growing on encrusted trees, can feed giant
herbivores such as bison, whilst boar forage around root
plates to find food. Rotten logs provide the perfect
maisonettes for shrews; hollow trunks for dormice. As a

result, martens and weasels find such fallen log haunts ideal hunting environments. Camera traps in Białowieża have shown that lynx use fallen logs to facilitate movement, allowing them to glide even more silently, and without hindrance, across the forest floor. Even beaver dams, built in woodlands rich in decay, are made largely of finer pieces of deadwood.

The importance of deadwood, however, stretches far beyond merely providing homes for a vast array of creatures and, in doing so, a veritable restaurant for predatory animals from wasps to woodpeckers and martens. 'Nurse' logs in wet woodlands are those where saplings burst from decaying, fallen logs – saplings that could not grow straight through standing water, but can grow from the petri-dish of fallen timber. Large fallen trees also soak up and sequester water, as well as locking away carbon, becoming, over time, reservoirs of their own within woodlands. Large dead trees, left alone on steep slopes, can play a vital role in stabilising those environments; anchoring soil in place and even resisting avalanches. Like the roots of living trees, dead trees continue to hold rocks in place, or prevent their sliding downhill. Yet for all these ecosystem functions, dead trees, lying prone and apparently beyond repair upon the forest floor, serve one utterly vital function; on land, it is arguably the most important ecosystem service of all.

The fallen boughs of mighty trees, lying prone in ancient or unmanaged woodlands, allow and protect the next generation of trees to grow. During their study in Białowieża Forest, Bobiec and his team closely studied one apparent ecological paradox in particular. Białowieża had *higher* densities of deer than many, adjacent managed forests, even though the deer in Białowieża are hunted both by wolf and lynx. In spite of that, those foresters managing adjacent forests had to regularly cull deer, or fence woodland, to protect new trees. By contrast, in Białowieża, no deer protection was needed for its ranks of new trees to flourish. The answer lies not only in wolves or other predators – but also in dead trees themselves.

Fallen boughs, logs, or other woody material serve as invaluable nurseries for future generations of trees by protecting them from browsing or grazing animals. They act as natural miniature barriers with some species, especially pine or yew, providing a bristled enclosure within which young trees can prosper. In a truly ancient woodland, complex networks of dozens of fallen trees – perhaps a stand of a similar age, which has fallen at once through gales or collective collapse of their roots – create larger nurseries, some as large as a square kilometre. If you walk through Poland's Białowieża today, you will therefore encounter enormously dense stands of young hazel, the haunt of marsh tits and hazel grouse, and groves of young saplings, trilling with wood warblers, in a woodland that is still the home of giant grazing animals and thousands of red and roe deer. It is, in truth, the trees here that protect one another, as, even in death, their protective, collective life force remains.

Dead trees are one of the most important ecosystem engineers of all; the culmination of hundreds of years of biodiversity which, even in passing, fuels and protects the next generation, and even in dying, grows ever richer in life. And so, this brings us back to how we might rewild our woodlands – and whether we should, in truth, plant and closely steward our native trees, and woodlands, in the future. As much as any wolf, lynx or beaver, any whale, boar or eagle, trees are designed to be wild – to form communities and interdependencies over time; to compete and yet protect one another; to create the conditions for other trees, or other kinds of woodland, to prosper. Trees have been doing this for millions of years – with enormous success. In doing so, they have stabilised our climate, provided for our diet and our homes, and supported many of our other wild creatures against all the odds; in spite of all the destruction we have wrought.

Should we, then, plant trees? Where necessary, then we must. But it is worth remembering, first, quite how well trees can run the world by themselves. Long before us, long after us, they can sustain and self-govern; creating and rejuvenating ecosystems through the simple act of living and the even greater benefits of dying. As we look to the ecosystems of the future, it is then perhaps wise to remember that trees are active wild forces, not passive monuments. It is worth remembering that trees are supposed to be wild.

CHAPTER EIGHT

Lynx and Wolves

British people, or at least a small proportion of them, currently live beside the greatest living hunter on Earth – larger and more powerful than a polar bear. We not only tolerate but laud the orca's return to the waters of Shetland and the north-western coast of Scotland. On land, however, the story is very different. The largest carnivore we tolerate, and today less so than we have in decades, is a moderate-sized black and white member of the weasel family: the badger. The prospect of living with larger carnivores excites some, terrifies others, and leaves few indifferent.

And this is why most advocates of ecological restoration believe that if this deadlock is to be broken, then of the three large carnivores native to Britain – the lynx, the wolf, and the brown bear – the lynx, impossibly shy and secretive, a creature of dense woodland, is the most likely candidate for a return to our shores. Ironically, we know almost nothing about it. To learn more, we need to turn to the intrepid scientists of an equally little-known country – Belarus.

'Where lynxes prevail, foxes will fail'. The sentence struck me for its wit as much as its content; scientific papers from Belarus are not always noted for the quality of their humour. But then the zoologists studying lynx in the Naliboki Forest of Belarus are not ordinary scientists either. They are field naturalists of extraordinary acuity. They can achieve, on a yearly basis, what I can only dream of – they can locate the dens of Eurasian lynx, and wolves, deep within a forest.

Even for professional nature guides, in countries like Finland or Poland with good populations of lynx, these mid-sized cats are as ghosts. I know professional guides in Poland's Białowieża Forest who, living beside lynx every day, have never seen one. And in the Bükk Hills of Hungary, in 2017, I joined this club of 'non-lynxers' in one of the strangest encounters of my life. I was in a small clearing within the forest, watching butterflies, when a terrified roe deer shot across the long-grass meadow and back into the forest. Here, I could see that it was frantically fleeing from *something*. Something that, given the blank look in the deer's eyes, did not want to make friends. The deer vanished seconds later into the deep undergrowth. I ventured in, sometime later, but I never saw the deer. Perhaps it had escaped. And I felt at least fairly sure that whatever its fate, I had, right in front of my eyes, been out-lynxed. Was it a lynx? Was it not? I will never know. And for some people, even those living beside this animal, that rather encapsulates the nature of the beast.

Lynx are best described, ecologically, as 'mid-cats'. They are not lions or pumas or jaguars, but instead fill the next niche down. Lynx are, as we see them, timid animals – except, that is, when it comes to their superpower: the ambush of their main prey, which in most of Europe is the roe deer. Whereas wolves are generalists – both in terms of the range of landscapes they occupy and the prey they hunt – lynx are specialists. They ambush roe deer, generally within mature woodland or at the edge, using camouflage, a short stalk to bring them within metres of a strike, and then a devastating burst of speed. A recent, remarkable camera-trapped video from Finland shows a roe deer at a feeding station, and a lynx creeping up in its blind spot. The lynx bursts forwards onto the deer's back, the deer buckles and then charges away – with the lynx still very much attached to its back. As the deer blunders into the woodland, and out of the camera's view, we see the lynx flip underneath the deer to seize its throat between two lethal pairs of canine teeth. This is a what a modern European big cat hunt looks like; a hunt that, except for such moments as these, remains hidden from view within the confines of our continent's larger woodlands. But in the past few years, more detailed studies of lynx, especially of those in Belarus, have revealed a more complex and surprising animal.

In Belarus, Vadim Sidorovich has studied lynx since the mid 1990s. Using thousands of hours of tracking, and camera traps, as well as more conventional telemetry, coupled with an amazing field instinct for finding dens that has grown more acute over time, Vadim and his team have invested more than 20 years now in the study of Europe's ghost-cat. Rather than seeking an academic approach to such a shy and complex animal, the team have adopted a 'non-academic research style', providing proofs through field documentation and actual observation, rather than the mathematical simulation which has, in recent years, come to replace some ecologists' connection with the field, and the actualities of animal behaviour. Their results have been fascinating, and they also show us how easily this silent hunter could, without interfering with us, live beside us once again in the future.

The team have found that the range of a lynx varies enormously depending on the animal, its size, age and position. Young males may occupy territories of only 10–30km^2, but large territorial males can wander across far larger tracts of woodland, closer to 200km^2 in size. Two to four females will live within an adult male's home range, but if the lynx population grows, males will overlap more often, and, under ideal conditions for the lynx, you may have as many as three home ranges squeezed within 100km^2. By way of comparison, the New Forest National Park – which, were it not for too many roads, dog-walkers and the relentless disturbance of a place where disturbance is legal and legitimised, would be eminently suitable for lynx – is around 560km^2 in size.

Once born, lynx kittens are hidden well by their mothers; Sidorovich and his team usually find dens under fallen boughs or large log piles, or within old badger setts, abandoned beaver burrow networks, and even old abandoned wolf burrows. The most important factors determining the location of a lynx den appear to be protection from rainfall, from mosquitos, and from other predators, particularly wolves. The interplay of such species has been largely forgotten across most of Europe, and indeed only in countries like Belarus, where 80 per cent of the land still operates under predominantly natural processes, can we get some glimpse of the complex sharing arrangements that once took place in Britain more than a millennium ago.

During denning, the lynx mother often pursues a more varied diet of smaller prey – including red squirrels, mallard ducks and young hares. These she can readily carry back to her kittens. During this kitten-rearing time, the team have discovered that rather than a successful jump from a hidden place, the lynx begins in an ambuscade before a short stalk or sudden chase to capture its prey. All of this takes place against a fearful backdrop of fragile kittens; kittens that could easily go missing or be harmed by predators or weather.

When lynx kittens are small, the mother will move them as infrequently as she can, often keeping them in one secure site for as long as two months. But often, chance will intervene and force her paw; if human disturbance or wolf pressure in an area of forest increases, the mother must move her kittens as often as once a week. Often, she must then hunt with the kittens in tow; a tough challenge indeed. As the kittens grow older, the mother will often begin to tackle larger prey: deer, or grouse, from time to time. At times, there is evidence coming together to suggest that the adult male in an area will team up with the mother to hunt. In this case, the team have observed that prey is brought back more regularly to the kittens – and the success rate would appear to be greater; a trait observed widely in more readily watchable species, such as African lions, where pincer and 'decoy' movements are both common during the hunt.

During the snowy months, the life of lynx kittens becomes hard indeed. Tracking suggests the mother can move them for four to six days at a time, over distances as long as 70km in the snow. During this time, the female once again extends the menu, catching not only red squirrels and grouse, but also sometimes young beavers or wild boar. Based on the behaviour of leopards, one imagines that the female must quickly grab a juicy piglet, and make off with it into the forest, where the short-sighted boar cannot pursue her. A struggle with an adult boar, especially a male, would likely not end well for the lynx. Whilst over time the kittens will leave their mother to find food for themselves, it would appear that this process is 'softer' with lynx than in other species. The Belarussian team believe that young females can, for a year after birth, remain close to the mother, rather than being forced immediately out. The adult male, meanwhile, has an entirely different set of behaviours.

The life of an adult male lynx is geared heavily towards patrolling, scent-marking and territory maintenance – much like any mid-cat or big cat, notably lions. During the winter months, males wander along well-defined forest tracks or routes that they know well, often stopping as often as every 100m to scent-mark them. Male lynx, it seems, are deeply conservative animals. They meticulously pass, patrol and re-pass 40km^2 or so of real estate, with a marked degree of consistency. Like bobcats and Canadian lynx, male Eurasian lynx call frequently from elevated spots. However, this behaviour soon becomes more dramatic. Male lynx can climb up to 30m, particularly in pines, to emit these calls. The Belarussian team have found scratch-marks this high, but it seems that male lynx only go to this effort if the population of competitors is high. In other words, he who climbs highest is perhaps most likely to attract females, and project his voice over the longest distance. Like most cats, it can also be surmised that lynx escape their own predators, such as larger wolf packs, by shooting up trees.

For a male lynx to successfully attract and retain females, the heart of his territory must incorporate top-notch hidey-holes, especially dense thickets and tunnels, where the female feels safe not only to mate, but also to leave her kittens. Finally, the male is also a creature of habit in the way that he hunts, often having favourite ambush points that he will use time and again, rather than hunting randomly across the landscape.

In recent years, however, the researchers have established that lynx may not be as antisocial or solitary as the scientific literature generally depicts. With an amazing degree of field knowledge now behind them, the Belarussian team made a particular study of a female, Hanna, and the adult male, Kazimir, with whom she was paired, and another pair, known as Bazyl and Bazylikha. They found that pair bonds between the animals could be strong indeed, keeping them together when the female didn't have kittens to care for. Cooperative living throughout the cold season now looks more and more likely. In addition, the team have found tracks of up to four lynx – a female, her kittens and a sub-adult – all walking and hunting together, and, at other times, Kazimir walking with Hanna and her two kittens.

Whilst all of this detail is fascinating and much of it unknown to the British public, the Belarussian studies paint the picture of a cat as dependable in its daily habits as the domestic cats with which we are far more familiar. It is clear that lynx need large areas, use these areas non-randomly and with a high degree of selectiveness, do not like their kittens being disturbed and require prey to come to them, or past them, at favoured ambush spots. This all provides British conservationists with valuable information as to where the lynx, if reintroduced back into our country – as many conservation advocates now propose – could thrive.

There seems little doubt that finding food within our larger woodlands, especially in Wales or northern Scotland, or indeed the larger forestry plantations of Kielder, Galloway or Tayside, would pose little problem – for these are places where squirrels, roe deer and mid-sized birds are not in short supply. Were lynx able to inhabit, over time, quieter areas of the Forest of Dean or Wye Valley, home now to sizeable populations of wild boar, there is little doubt that their ability to snatch piglets would be to their advantage too.

Far more has come of the Belarussian study, however, than the ecological basics of how lynx use and occupy a wooded landscape. One of the more surprising elements has been the degree to which lynx control other predators and 'regulators' in the landscape – most of all, the fox. During their study of radio-tagged foxes in the forests of Paazierrie and Naliboki, the team found that more than half of the animals tagged ended up being killed by either lynx or wolves. During the denning periods of the lynx, especially, the necessity to kill foxes increases, as foxes can predate small lynx kittens. The team found that as wolves and lynx increased back to natural saturation point in the Belarussian forests, a situation still incredibly rare in Europe, the fox became something of an endangered species. It was found that lynx and wolves dealt with foxes in a different manner. Wolves excavate out a fox, whilst another wolf, lying in wait, pursues and kills it. Lynx, however, eradicate all the red fox cubs that they can get their paws on. Several years on, the fox has entirely changed its behaviour in the Naliboki Forest, spending 80 per cent of its time in grassy clearings. This alone renders the lynx a powerful ecosystem engineer because, as any conservationist or indeed gamekeeper will attest, the impact of foxes, in the absence of depredation, can be considerable.

Foxes, being mesopredators, have a broad diet, as evidenced by their ability to live within and indeed thrive in British cities. They are, however, regular predators of birds' nests and chicks. It has long been argued that 'fox control' is a necessary part of conservation, especially for those ground-nesting orders, such as wading birds, cranes or even raptors such as harriers or merlins – and ecologically speaking, this is not untrue. What has been forgotten, however, is that if lynx and wolves both dominate an environment, foxes never reach saturation point. This only happens in prey-rich, large-scale environments, where lynx, especially, can attain relatively high populations. The ability of lynx to regulate smaller predators, and therefore reduce the impact of species like the fox, which in turn can exert excessive influence on smaller species such as ground-nesting birds, is one of the greatest powers exercised by lynx in their woodland environment. But their impacts upon deer are also considerable.

On average, when roe deer proliferate, they are the favoured prey of the lynx. If roe deer are present in good numbers, they can comprise up to 90 per cent of a lynx's diet. The Belarussians found that a male will average one roe deer every six days; a female with dependent young kills a roe deer every four to five days. Other studies have sought to quantify to what degree such activity affects the deer within Europe's larger woodlands, and one study achieved this by looking at what happened in eastern Poland, where lynx were hunted out of the landscape. Here, it was found that when lynx were removed from Polish forests by ourselves, the roe deer population skyrocketed. Though by no means the only factor in deer control (weather and forage reduction, and consequent starvation, being two others), the return of lynx to Poland's forests saw a concomitant drop in the density of deer. In a Swedish study, Henrik Andrén concluded, after detailed research into kill rates by lynx on roe deer, and allowing for other factors (including fox predation of fawns, food and weather-related factors) that the lynx is a 'very efficient predator on roe deer, capable of reducing roe populations to low levels'. This has enormous ecological significance when it comes to considering lynx reintroduction to Britain – where we face an overbrowsing situation by deer that is perhaps the worst on the continent.

In Britain, roe deer currently number half a million, of which around 150,000 animals live in England, and 350,000 in Scotland. Whilst deer have not been naturally predated by lynx for centuries, they have, nonetheless, increased further in recent years for a number of other reasons. Mild winters are thought to have played a part, and because so many of our woodlands border arable farmland, deer have been able to find ever more reliable winter food sources. The lack of brutal, long winters, familiar to our Victorian writers and chroniclers, has meant failure to weed out the sick and the weak. Roe deer, like other deer species – such as the alien muntjac, or the red deer, especially in Scotland – have become a rampant force in our woodlands. The effects of too many fearless deer, in a woodland, is not something readily observed.

The greatest effect of deer, in contrast to digging animals such as boar (which do not eat saplings), or large grazing animals (which tend to pluck vegetation), is that their browsing can effectively prevent the growth of future generations of trees. This is generally achieved by intensively browsing, or clipping off, new shoots, young leaves, and small branches, during a stage in the tree's life where it is intensely vulnerable. This behaviour in deer is not seen in ecosystems where large predators remain, for the simple reason that browsing trees to death takes time – whereas a fearful group of deer, accustomed to being hunted by a large predator, are extremely unlikely to tackle any one tree, or sapling tree, for any sustained length of time, before moving on.

Therefore, in areas of Europe, especially woodlands in eastern Europe, where lynx, and sometimes wolves, conspire with more sustained human hunting pressure, a woodland is a very different affair to the gaping, open-floored forests we have become accustomed to in much of Britain today. In areas such as Białowieża, the ground flora is infinitely more developed – home to pristine carpets of flowers. The woodland edges of many Carpathian forests, in Hungary and Romania, are a riot of butterflies and their caterpillars, many of whom are feeding in the rich array of thorn, nettle, flowers and scrub left intact at the woodland edge. Bramble, which for all its toughness is readily digestible by deer, can be found far more readily at the woodland edge than in many British woodlands; feeding butterflies and providing invaluable nesting and sheltering sites for woodland-edge birds such as nightingales. New saplings, often prime candidates for overbrowsing by deer (including roe) here in the UK, are a considerably more common sight. They provide feeding grounds and song-posts for wood warblers and a whole suite of 'small tree' bird species. Some areas, especially in damp regions, grow dense with young bushes and thickets – the favoured home of nightingales, garden warblers or willow tits. In short, woodlands in the UK are now mown into a homogeny of bracken-covered floors; one of few plants unpalatable to deer. Often, this damage is subtle, and easily ignored – but only on visiting a English deer-fenced woodland, excluding deer from browsing within, does it become readily apparent the damage that deer do when left unchecked elsewhere. Our woodlands should be full of flowers, bramble, scrub, glades, 'open' sunlit floors and shady thickets. The return of the lynx would not solve all of these issues – it would, however, certainly help. And in promoting a richer and more varied woodland, the effects of the lynx are likely to be widespread, indeed, more so than has yet been studied. In fact, putting the lynx back into a relatively butterfly- or bird-poor forest in the UK is likely to have a greater effect than when lynx increase in Belarus, where so many species are, due to a lack of habitat degradation, already thriving.

Another surprising effect of lynx is its ability to enrich and transform woodland soils. Because lynx will generally not eat all of a deer carcass, feasting mainly on the freshest and best meat, the carcass is then left for a range of smaller meat-eaters to find. However, parts of some carcasses will remain, enriching the soil where they lie, and promoting the growth of new flowers and grasses.

There is one, final regulatory function the lynx provides – and this one almost beggars belief. Over many years of studying lynx and wolves side by side in Belarus, the team there found that contrary to all popular belief, and indeed much scientific lore, the lynx is often the regulator of the wolf – and not the other way around. In early studies, the scientists were curious to observe that wolves, upon smelling lynx, do not actively enter their den, or prey upon lynx kittens. Indeed, the only lynx found killed by another carnivore were male lynx killed by other males. Lynx, however, did not show the same deference to young wolves. In many years of study, the team found that lynx had killed wolf pups as old as six months, recording this as many as 10 times. In addition, lynx killed a pregnant female wolf. So whilst lynx are not known to tackle wolf packs, and would most likely lose, or die, if they did, they are perfectly capable of killing single wolves. Lynx do not eat those wolves that they kill: instead, this is known as 'interference competition', whereby organisms directly competing for resources may resort to predation, or aggression, to achieve their ends.

But the wolf's fear of the lynx goes further than its reluctance to enter a lynx den. Camera-trapping studies show a lynx crossing a dead log 'bridge'. Hours later, a wolf appears, but walks with extreme caution, fearfully inspecting the snowy lynx tracks with trepidation. Other camera traps have shown a pack of wolves approach a tree scent-marked by a male lynx. Rather than show signs of aggression, the wolves flatten their ears, adopting a clear posture of submission. In one startling camera-trap video, an eight-year-old lynx confronts an adult male wolf in the heart of Naliboki Forest. The lynx pins the wolf on its belly, biting it, and the wolf flees the scene, as the lynx returns to mark its scent-post. Later, another camera trap picks up the same, distinguishable but now gravely injured wolf. The lynx, it seems, may, in a truly pristine woodland environment, be the closest thing the wolf has to a natural predator. Indeed, in recent years, the team in Belarus have taken this supposition even further.

Since the recent increase of lynx in Naliboki, from 22 in 2013–14, to 60 by 2016–17, three major events have befallen the forest's population of wolves. Firstly, the number of litters and pups have declined markedly. Secondly, and potentially as a result, the immigration of new adults and packs into the forest has markedly increased in the winter months. Finally, come spring, there is significant *emigration* by wolves out of the forest. The team have theorised that, if confronted with high densities of lynx prior to their denning season, wolves may move out of woodland and into more open lands.

The implications of this for how Britain's ecology might once have functioned are profound. Just as Belarus is currently a country of 60 per cent native woodland cover, the rest being grasslands, farmlands, bogs and huge river systems like the Pripyat, so too it is thought that Britain, prior to Bronze-Age deforestation, would have had a similar amount of woodland cover, leaving around 20 per cent of scrub or scrub-grasslands, and 20 per cent of vast fenlands, prior to their drainage far later in our history. If lynx were dominant within our woodlands, this may well have meant that wolves could not operate well in such habitats, hunting instead in the margins, open land and floodplains where, as the more adaptable of the two carnivores, they would have thrived. It will be many decades before such dynamics are seen in Britain, perhaps even longer. Meanwhile, though, Europe's most pristine country, Belarus, and its interlocking carnivores, provides a fascinating reminder of how apex predators do not only regulate ecosystems, but also regulate one another. For now, however, the quest of those seeking ecological restoration is to see lynx here in Britain at all.

Unlike the wolf, the lynx leaves virtually no trace, neither in the environment, nor in British or in fact most European folklore. Indeed, before the Belarussian studies, few sustained and intimate insights into wild lynx populations existed anywhere in Europe, reminding us just how low-key the animal can be, even in countries home to hundreds of lynx. But there are exceptions to the rule. Whilst lynx attacks on people are unknown, and this fact is now widely accepted, lynx predation upon sheep, whilst generally low, to very low, can vary – and at the higher end, it can be problematic, albeit not catastrophic. The question is whether the damage wrought by lynx, occasionally, on some flocks of sheep, might justify forever banning such a graceful and unique animal from returning to our shores, or outweigh the other benefits – social, economic and ecological – that returning this animal would provide.

On examining the evidence of how lynx interact with livestock, and thus with the livelihoods of farmers, it is always best to take the worst example first, if only in the name of fairness to those who oppose their reintroduction. In this regard, the worst and only noticeable effects of the lynx on livestock in the whole of Europe, come from Norway. Here, sheep are moved in a very different way to the UK. Often they are shepherded through, and then left within, wild, wooded environments – rather than open hillsides, pens, fences or fields. The result, unsurprisingly, is that by shifting sheep directly into the prime habitat of the lynx, sheep predation becomes considerably more commonplace. In a study carried out between 1994 and 1999, 34 radio-collared lynx killed a total of 63 domestic sheep, across six grazing seasons, averaging 10 animals per season. It was found that male lynx were by far the main culprits in this matter, with a far lower rate of predation amongst females. Norway's lynx–sheep situation is, however, remarkable – insomuch as it does not occur elsewhere.

In the French Jura, where the landscape is more similar to many British uplands – with a harder demarcation between fields, some open grazing land, and forest – and sheep are not generally moved into wild woodland areas, perhaps a better field of reference exists for how lynx might adversely impact on livestock farms (indeed, there is little evidence from any group that lynx would impact adversely elsewhere). Here, clustered attacks within small areas were noted, these 'hotspots' often covering less than 4 per cent of the range of a lynx. Sheep taken here constituted around 3 per cent of lynx diet, and this was the highest percentage recorded across the studied area. The fact that the same, small hotspots of predation were identified each year, suggested two things to those involved in the study. First, that a relatively small number of lynx were involved in these attacks. Second, that particular features in the landscape may have aided ambush, or that particular husbandry techniques, in those areas, led to sheep triggering an ambush reaction from a lynx. In total, 80 per cent of studied flocks were never attacked during the study, meaning that the chances of a loss of livelihood among farmers was extremely low. The attacks were 'episodic and sporadic' – and in France, as in Germany, a compensation policy was adopted for when these took place. Currently, whilst the wolf raises far larger problems in farmed areas within Europe, in France, Germany and Switzerland, all sheep-farming countries living beside lynx, illegal persecution of lynx remains low, and farming remains viable.

It must be said that for all parties in Britain, especially famers, the *prospect* of living with lynx sounds far more dramatic, and radical, than what actually happens when lynx are back in the environment. Whilst a powerful steward of our woodlands, our foxes and our deer, and innately wonderful creatures in their own right, lynx are among the most ghostly of all animals, on a par with pangolins or Siberian tigers in terms of their public profile. That said, just because lynx are invisible, does not render them unable to dramatically improve local economies, or even effect a range of other outcomes too.

In Poland, lynx are broadly welcomed by foresters, because whilst they may scratch the odd tree, they save many more saplings and young trees from overbrowsing through the effects they exercise on roe deer populations. If lynx may be unwelcome for some UK sheep farmers, they would be considerably better news for arable farmers if hunting the woodlands adjoining those farms. By hunting deer at the woodland edge, lynx would reduce the impact of deer upon our crops; a multi-million pound 'damage industry' in Britain each year. What's more, the impact of lynx upon cattle farming would most likely be

either neutral, or positive. Lynx simply do not tackle
calves; they are not evolved to, and there is no evidence
that they do. Lynx do, however, displace badgers from
their dens, and will kill badger cubs if given the chance. In
doing so, they are likely to have some impact on badger
densities around farms, something which, for reasons of
bovine TB, may be broadly welcomed by farmers. And for
those ecologists concerned about certain species, such as
the wildcat, in the event that the lynx makes a return, the
visionary conservationist Roy Dennis makes the following
observation:

> When I hear people say that we cannot bring back the lynx
> for fear of putting paid to the wildcat, I wonder if they
> understand the functioning of ecosystems for wildlife
> conservation. I remember one winter riding through deep
> snow in a Carpathian forest and coming across a wildcat
> eating the remains of a roe deer under a hazel tree. I had
> earlier followed the footprints of a lynx along a forest track.
> My hosts, experts on large carnivores, knew exactly where I
> had seen the wildcat because they had seen [the lynx] with
> its kill in the snow several days earlier. To them, lynx and
> wildcat were both simply part of the wildlife community in
> the mountains of Romania.
>
> Instead of posing a threat to wildcat, the impact of
> lynx on fox and badger would undoubtedly, in my view,
> benefit the wildcat by reducing the numbers and ranging
> behaviour of its competitors. In the very long term, true
> recovery of wildcat may not be possible without restoring
> the lynx.

The final beneficiaries of returning the lynx, however,
would be the majority of people living within communities
local to lynx populations. As has been reliably documented
in the Harz mountains of Germany, lynx, through their
allure, mysticism and beauty, have the potential to draw

ecotourists to an area – even if those tourists *know*, deep down, there is little chance of actually seeing a lynx. The Harz mountains, indeed, act as a good region in which to study the practical and economic effects of lynx in the landscape. Here, lynx have been returned to the landscape for the first time in 200 years. The allure of seeing wild animals has seen £12.8 million invested in the local economy, a small and predominantly rural community comparable to many of our own. But only 12 per cent of those who visited, among respondents, said that in retrospect, they would not have travelled had they thought they wouldn't see a lynx. In other words, it is often the allure of lynx, the knowledge that they are there, sharing the landscape, the mere prospect of a glimpse that can, reliably, drive tourism and see millions invested in those areas where lynx are present. More than half of those respondents visiting the Harz cited lynx as a primary reason to visit, and the geographic pull of those visitors came from across the northern half of Germany, not only the immediate vicinity. In addition to walking in the presence of the lynx, Harz locals have cleverly monetised the potential of the animals. In addition to an enclosure where the public can see captive lynx, set for future release, there are also tracking tours and the inevitable gift shops.

The Harz is a large, forested area, capable of sustaining both tourist pressure and a viable, wild population of lynx – with enough space and solitude to ambush in peace, call, mate, raise kittens in a den, move said kittens and roam across the landscape at a 100km level. There are in Britain only a few regions in which this could viably happen, most of which are in Scotland. It is therefore important to accept that lynx will not single-handedly solve the British deer problem, nor will they, sadly, flourish in the road-bisected, dog-trodden small woodlands in much of southern England. But nor should we be too conservative as we think about how these cats could live beside us in other areas. Provided lynx have remote, prey-rich landscapes, and those landscapes are wooded and undisturbed, it appears they can flourish. And this leaves several areas of Britain eminently suitable for their return.

On the border between England and Wales, the combined woodlands of the Dean and the Wye, the latter linear but steep, dense, and slinking westwards for miles into Wales, may still be large enough for lynx, and are most certainly prey-rich enough for their needs, packed with roe, boar piglets, squirrels, ducks, rabbits and rodents. A relatively dense network of roads bisects the area, and tourism and walking pressure is high. This area might be able to sustain lynx, but it is unlikely that this is where anyone would begin when contemplating such a bold move in conservation as the restoration of Britain's largest contemporary cat. Indeed, it is in the darker, steeper, denser and less accessible woodlands of somewhere like Afan Forest, in South Wales, which, by joined moorlands and plantations, extends north-westwards to Llandovery and then into Snowdonia, may provide a larger, quieter corridor for lynx to hunt and move. But with up to 80 per cent of the Welsh land mass given over to sheep, it is somewhat unlikely that the lynx will land here first.

In northern England, significant tracts of quiet, dense woodland also exist – many of these unnatural, being spruce plantations of otherwise poor wildlife value, relative to their enormous size. However, the combined acreage of Kielder, extending northwards into Hawick Forest and then, in a largely unbroken web of forests, west, eventually into the enormous Galloway Forest Park, would, most certainly provide the acreage, space, feeding and breeding requirements for the lynx. But it is further north, in areas of Scotland already heavily given over to woodland, reforesting and rewilding, where perhaps the soft paws of the lynx might first tread this land for the first time in centuries.

Here, in areas like the Cairngorms National Park, rich mosaics of ancient Caledonian pine and forestry are already well protected, continuous, growing, and home to other shy giants of the woodland, including most of Britain's remaining capercaillies. Furthermore, such habitat extends, via remote and narrow glens, into the woodlands of Deeside, to the east, the woodlands of Forres and Fochabers, to the north, and, from Inverness, into the quiet and remote woodlands that run the length of Loch Ness to Fort Augustus, Fort William and beyond. Once established in such an area, it is likely that the lynx, accustomed to becoming the ghost of Europe's woodlands, would be able to thrive – finding dens and hunting sites away from people, hiding away in the many steep or little-trodden hillside woodlands of the glens. This part of the north-east Highlands is striking not only for the degree of woodland cover, wildness and seclusion, but also for the fact that major roads (which could cause damage to such a low-density, slow-breeding species, as has happened with the Iberian lynx) are conspicuous by their absence. Additionally, huge tracts of habitat within the Cairngorms, south or north of Loch Ness, and around Deeside, are home to just one major road. In this large, low-intensity and heavily wooded landscape, it is widely thought by those that support ecological restoration that the lynx – fulfilled by the hundreds of thousands of roe deer, as well as red squirrels, mountain hares, ducks and some grouse – would thrive. And from just two or three reintroductions, over the course of a few years, this magnificent cat could once again be an established inhabitant within the woodlands of our own island.

But there is of course another. One more powerful. One whose strength lies in the pack. In mainland Europe it now wanders through farms and villages by night, and hunts quietly in the woods and mountains beside them. In Britain, to even speak its name is to conjure howls. We are not ready for it yet. But if one day the wolf returns, and we can, in some parts of our country, learn to live beside it – it would change the British landscape forever.

It is perhaps one of the great ecological ironies that we cultivated man's best friend from what for many would still be seen as man's worst enemy, the wolf. Our battle to de-wolf Britain began early on in our history, and with an increasing, organised ferocity until the animal's eventual extinction. The prominent ecological historian Derek Yalden estimated that there would have been around 6,600 wolves present in Britain prior to their persecution, based upon the range sizes currently found in Poland's Białowieża Forest, although even these have been depressed through hunting. Yalden estimated a similar number of lynx.

Whilst there is little trace of how the lynx vanished from Britain, with early deforestation and removal of its prey-base, due to human hunting, deemed as likely as direct persecution, the wolf was a more adaptable animal, and hunting played a far bigger role in its demise than habitat loss. Indeed, whilst wolves do inhabit forests, they are arguably just as successful in mixed, semi-open landscapes. As this is written, wolves are thriving not only in the deep, dense forests of places like Romania, but also in the Alpine farmland edges of Italy, the farms of eastern Poland, the open meadows and wood pastures of Portugal, the treeless steppes of European Russia and in the dwarf-birch clad taiga of Finland. So even as we deforested Britain, it would be our hunt for the wolf – deep and direct – that precipitated its demise.

As early as Anglo-Saxon times, records were made of King Edgar demanding, from Wales, the skins of wolves as a tribute, at a time when wolves were especially common on the wooded borders of England and Wales. Wolf bones have been found widely in Anglo-Saxon burials, and places like Woolpit, in Suffolk, named as early as the writing of the Domesday Book, are thought to have referred to 'wolf-pits'.

At this time, wolves were still abundant, especially in the forested and upland areas of Britain, but their systematised hunting began early on. Some criminals, rather than being put to death, were required to provide wolf tongues instead – one imagines that more than a few may have perished attempting such a feat. The Anglo-Saxon chronicles, however, document a far more organised and well-regimented hunting operation. A closed season of wolf-hunting for the nobility would begin in the 'wolf monath', or month, of January, continuing into late March. This was Britain's wolf-cubbing season. At this time, hunters would sometimes hunt wolves actively with spears and dogs, but more often placed pitch within woods frequented by wolves. However, in dry summers they would also seek out and kill wolf cubs. It is often forgotten that wolves were not only hunted because they competed with the nobility for deer, or threatened livestock, or us. Wolves were also hunted for their pelts, which were deemed of greatest quality during the denning months.

By Norman times, servants to the kings, until 1152, were tasked with ridding the land of wolves, which, even at this time, appeared to have become rare in southern England. As late as 1212, a bounty of 12 shillings was paid out for a wolf killed in Hampshire which, at that time, would have been a very significant payment for someone working on the land. Like so much of British wildlife extermination, the prospect of significant sums of money, especially for the wolf, would have motivated those in every region of England to hunt these animals down. That said, by this stage, there were already few wolves left in the

predominantly forest-free areas of southern England, and
so one imagines that any appointed wolf-hunter may have
become more and more professionally disappointed as
time went on. In the late 1100s, there was still a healthy
wolf population in Wales, as Giraldus Cambrensis recounts
wolves emerging from the woodlands to eat the putrid
corpses of battle left from when Henry II attacked the
country in 1165.

By 1281, the death knell would come for remaining
English wolves, as Edward I ordered the extermination of
all the wolves in England. At this time, Gloucestershire,
Herefordshire, Shropshire and Worcestershire were, even as
they are to a lesser degree today, more heavily wooded, and
less populated, than southern England at this time, and it
was here that Edward's wolf-hunters turned their attention.
If one man in history has been reliably documented to
have removed more wolves than any other, it was Sir Peter
Corbet, on behest of Edward I – a legendary hunter of the
time. For nine years, Corbet and his pack of hounds
traversed what would at that time have been vast woodlands
– and by 1290, when one of the last wolves was killed in
the Forest of Dean, the wolf in southern England became
a thing of a past; consigned to myth. Wolves would persist
in England for longer, in what are still wilder areas of our
country, like the Forest of Bowland, in Lancashire. But
further laws would be enacted to seal the wolf's demise,
and they are thought to have been removed from England,
or relict refugees, by the reign of Henry VII in the
late 1400s.

It is a reminder that Scotland was de-wilded far later
than England, that wolves would persist here for at least
another three centuries, perhaps longer. Indeed, wolves still
thrived in Scotland as late as the 1500s, when Mary, Queen
of Scots, hunted for them in 1563. By the latter end of the
century, large clearances of forest across the southern
uplands would drive wolves into ever greater conflict with

our expanding livestock, and farming. In 1577, James I made wolf-hunts compulsory and from here on, within a century, the wolf would all but vanish from the Scottish landscape. Whilst one famous record states the last wolf was slain in 1743, it is deemed more likely that wolves, being able to persist at low density, cleverly evading people and haunting remote and marginal areas when hunted fiercely, may have persisted in Scotland until the 1790s, or, indeed, even the nineteenth century. But they persist no more. Britain has no wolves – nor will they ever return except at our behest.

Given the extraordinary efforts our ancestors made to rid us of wolves, it is perhaps wise to at least begin with the notion that some logic dictated their actions. Albeit hunted out during a period of burnings, witch-trials and public hangings, there was reason, at that time, for removing wolves. They competed against those nobles who kept deer on their land, at one end of the wealth spectrum, but they also hunted sheep and sometimes, very young cattle. The more we deforested, the more they came into contact with us. That said, our determination to remove every single wolf, even from areas like the Welsh Marches where, it seemed, they stayed largely away from us and within dense woodlands, suggests that wolves occupied our imaginations, and fears, far more than they actually threatened the average person, or farm, on a daily basis. Indeed, by the time wolf edicts were issued, the species was already rare in the more populated parts of Britain, and was clearly, even at that time, a refugee of forests where, no doubt, it had learned to keep well away from people. That said, in a time before technology, before subsidies, before modern farming methods, wolves would have threatened farms, reduced productivity and some have argued that the burgeoning British wool industry – now virtually extinct, but hugely lucrative at the time – may owe a lot of its success to the removal of the last wolves from England and Wales.

Today, Britain has become one of the last European countries not to be repopulated with wolves. It is not that wolves have been consciously put back into many neighbouring European countries. It is that, in the absence of sustained and intensive persecution, in some cases, they have, rather, crept back in. Right now, the nearest wild wolves to Britain can be found in the Veluwe area of the Netherlands, and, across the Channel, as many as 530 adult wolves, in over 40 packs, are now living in France, of which most live in the heavily forested areas of the Jura. The wolf is on our doorstep – but should we let it in? The first question to ask, then, might be what wolves do – and what benefits they might bring if we could ever learn to live beside them once more.

Wolves are, among predatory species, the ultra-engineers of a landscape. More powerful than lynx, they shape the behavioural dynamics of many other species, both smaller predators and prey. Their effects upon ecosystems are well documented, and those effects grow as wolves reach saturation point within a landscape. Whilst a lone pair of wolves may have little effect upon a landscape, a landscape-level population of wolves, consisting of multiple packs can, given time, and the absence of persecution, transform that landscape completely. In this regard, the single most important thing wolves do is plant trees.

The primary effect of wolves in a landscape is upon the ungulates, especially deer, but also elk – another lost giant of our shores – which, if left to its own devices, is capable of significantly denuding a landscape. Anyone doubting the impact of red deer in the Scottish Highlands, for example, has only to visit the heather-smooth hills here and wonder where the trees have gone, and why they aren't growing back, to understand that deer can both denude – and maintain a denuded state – within landscapes where they are not actively hunted. The same is true of elk – larger even than the red deer, and a good species for maintaining open glades and trimming back willow, aspen and birch scrub, but, in large numbers, unchecked by predators, a mowing force within a landscape. And in this regard, whilst to some readers a familiar tale, it is still worth recounting what happened to Yellowstone National Park, in the USA, its landscape, and its animals when wolves were reintroduced.

By the 1920s, whilst otherwise well protected as a wilderness, Yellowstone had been cleared of its wolves, as indeed had occurred, sooner or later, in most of the western world. Because Yellowstone was studied intensively after the departure of its wolves –an intensity that was never applied to Britain's de-wolfing in the 1600s – even early on the alarm was sounded as to what was happening to the park. The landscape was visibly degraded. Huge herds of wapiti (*Cervus canadensis*), a large deer akin to our own red deer, roamed freely, without fear. Plants were dying off, and new trees failing to grow as the deer munched and nibbled unchecked.

It was discernible that the ecosystem was incomplete, and in 1995, after having carefully confined Canadian wolves in a large enclosure within the park, into which was snuck a tasty wapiti carcass, so as to acclimatise the animals to their new range – the grey wolf was once again released into the wild. From here on, one female, '#9', gave birth to the park's first pups in the wild. Since then, numbers and packs have increased; as of 1 April 2019, 61 wolves roam Yellowstone in eight packs, although total wolf numbers now fluctuate between around 80 and 110. The wolf is now back in sufficient numbers, and density, that scientists have been able to measure its effect upon the landscape.

In the years that wolves had been absent from the park, wapiti, and other deer, had had things all their own way. Whilst bears and mountain lions exercise some impact upon wapiti populations, wolves are the main regulator of wapiti numbers. In this regard, wolves perform two key functions: they cull, and they instil fear. On their return to Yellowstone, rising wolf numbers have killed twice the number of wapiti that was predicted. However, as the wapiti now recognises the wolf as a predator, an even more vital new factor is injected into the ecosystem – and that is the factor of fear. For the first time in decades, constantly wary and fearful of predation, the wapiti move around. And this transforms the ecosystem entirely.

Much as in the Highlands of Scotland, willow and aspen are two of the most vital trees in Yellowstone, if given a chance to grow. They line waterways, provide rich forage for beavers, provide shelter and nest sites for birds, and feeding sites, as rich invertebrate banks, for many other species. They are also the prime forage of the wapiti. By constantly moving wapiti through the landscape, as much as hunting them, wolves have slowly begun to transform the very nature of the wooded landscape in Yellowstone. And one fact is perhaps most surprising of all. There are now *three times* the number of wapiti in Yellowstone as in the 1960s, when wolves were absent. And yet, in spite of this, there has been an immense resurgence in willow. The key factor is not predation – but fear. Just as sparrowhawks change bullfinch behaviour by keeping these fat finches secretive, obscure and bound to scrubland habitat, so wolf pressure means that wapiti can nibble and trim willow, and help maintain open spaces for other species to thrive – but they can never linger long enough to browse these species into oblivion. Furthermore, under pressure from wolves, herds of wapiti break into smaller groups, to avoid detection, or lower their profile to wolves. Now, studies within Yellowstone show that in the past decade alone, millions of new aspen and willow saplings, which, under intense browsing pressure from wapiti, would simply have been denuded or maimed beyond repair, have begun to grow once again.

As the landscape has become more wooded, especially in rich alluvial areas like river valleys where willow naturally takes root the fastest, the wolf has turned out to be the saviour of another of our ecosystem heroes – the beaver. In this regard, another fascinating experiment was conducted in Yellowstone. It was shown that whilst those young trees coppiced or chewed by beavers turned into new, healthy stands of willow, those that were, additionally, browsed by wapiti, had only a 6 per cent chance of recovery after a few seasons, compared to an 84 per cent chance if coppiced by beavers – but *not* browsed afterwards by wapiti. In other words, whilst willow and aspen are, as in Europe, perfectly adapted to thrive under the stewardship of beavers, it requires another ecosystem architect – the wolf – to protect them from overbrowsing by deer. In short, wolves in Yellowstone are creating healthier trees by keeping wapiti fearful and mobile – and in doing so they are helping beavers too.

Across Yellowstone in the past two decades, watercourses have become, once again, diversely wooded and shrubby – leading to a boom in the small mammals, birds and insects that can thrive there. And with beavers discovering a new abundant food source throughout the winter, all their extraordinary effects upon the landscape, as noted earlier in this book, have also come to pass. In this regard, the effect of wolf, in enhancing the role of the beaver, has percolated down through every arena of life in Yellowstone, from its spawning fish to nesting birds and wetland floral diversity. But the wolf effect does not end there.

In ecosystems devoid of successful pack predators like wolves, even otherwise healthy or functional ones, such as areas of the Caledonian Forest in Scotland, one of the many forgotten ingredients of biodiversity is the presence of carrion within the landscape. In some parts of Europe, like the hills of Spain, carrion remains a part of the farmland ecosystem, insomuch as cattle die and can still legally be left on the hills – one of the reasons for the extraordinary abundance of vultures, and other large birds of prey, in areas of central Spain such as Extremadura. But with wolves in the game, the presence of carrion within a landscape transforms its ecology dramatically. Prior to the reintroduction of wolves to Yellowstone, harsh winters and starvation were the main factors acting upon the wapiti population. But these tend to produce seasonal, not year-round casualties. In the presence of wolves, a landscape is fed with carcasses all year-round.

American Indian legends that ravens follow wolves (something that has recently been observed in Poland's Białowieża Forest as well) soon came true, as these intelligent corvids came to associate wolf packs with the presence of food. Eagles, magpies, coyotes and bears all benefited from wolf kills – but so did at least 57 species of beetle dependent upon carrion; beetles that in turn feed small mammals, and insectivorous birds. And in looking at Yellowstone, it is of little surprise that many ecologists now regard the wolf as one of few species capable of saving the remaining large-scale, rural landscapes of Britain – most of all, huge swathes of the Scottish Highlands.

Whilst the lynx may prove a powerful force in regulating our woodlands and the smaller browsing animals that denude them, the ecological effect of wolves is considerably more powerful. Able to predate red deer – a very similar species to wapiti – and indeed proven to do so across Europe, wolves would both limit the numbers of red deer currently denuding vast tracts of northern Scotland and instil within them an entirely new behaviour, one which even deer-stalking cannot replicate: *fear*, and the continual need to move. It is widely held that the presence of wolves alone, even in relatively small numbers, and carefully regulated, if not regularly culled, would still transform the northern Scottish landscape over a matter of decades. The greatest single benefit would be the regeneration of huge areas of native woodland, especially pioneer species such as willow, aspen and birch, and the consequent removal of millions of tonnes of carbon from the atmosphere. Trees are also the most effective land cover for maintaining water quality; natural sponges that slowly sequester water, helping it soak into the soil and filtering rainfall into our rivers and streams. The economics and ecological sense of natural mass arboreal regeneration alone make wolf reintroduction a credible consideration for discussion in the years to come.

But the effects would extend far beyond this. In creating a landscape of fear within northern Scotland's wild areas, wolves would dramatically enhance its waterways and waterside trees. Their presence would improve life for Scotland's golden and white-tailed eagles, but at other times, it would be their ability to terrify, kill and reduce orders of smaller predators that would become of great importance – including to some of our most beloved birds.

It has been observed from Yellowstone, where wolves greatly reduce numbers of coyotes, to eastern Europe, where wolves in countries like Belarus remove large numbers of foxes, that where you have a large predator, the mid-sized predators of similar kind, like the fox, tend not to fare so well. In a Scotland once again haunted by lynx and wolves, the population of the fox, and its impact upon other species, especially ground-nesting birds, would most likely be heavily reduced. As many populations of species like curlews, now free-falling in numbers across the UK, depend not only upon good areas of food-rich habitat but also relatively low levels of ground predation, the effect of wolves (which rarely bother to eat birds' eggs or chase smaller birds) would likely provide protection to those species for whom the fox is a persistent and serious predator, as surely as the presence of goshawks reduces densities of crows and magpies, alters their behaviour and provides relative safe-zones for smaller nesting birds.

It has been widely observed that Britain is beset by mid-level predators, or mesopredators, many of which have, in aggregate, a powerful effect upon the eggs and nestlings of birds. From the pine marten's impact upon capercaillies and goldeneyes in northern Scotland to the impact of the badger upon the nests of species like wood warbler and woodcock, or upon the humble hedgehog, Britain's mesopredators have, for a long time, been able to rule an ecosystem where they are not meant to be in charge. For centuries, another apex hunter, humanity, has supressed these orders, though often to protect our own interests, whether farming or hunting, rather than the longer-term interests of the ecosystem. Now we have a situation where in some places foxes are hunted, whereas in others they are not. In some areas, sudden badger culls spring into being, often for spurious reasons that might be addressed by vaccination. In other areas, weasels or stoats may be removed. But there is no consistent, regulatory force to these occasional human depredations on a par with the continuous, 24-7 threat of a lynx pair, or wolf pack, to regulate mesopredators continuously, keeping them moving, killing some and shifting others out of entire habitats, or to the very margins. And that brings us back, again, to the brilliant Belarussian naturalists of Naliboki Forest.

In Scotland, studies have found that badgers average between one and eight individuals per *one* square kilometre. Anyone who has spent any time in the Caledonian Forest will know that whilst crepuscular and elusive, badgers are common animals, as indeed they are across many other wooded parts of the UK. By contrast, in Naliboki Forest, even now that the badger has recovered from the impact of alien racoon dogs, the population sits at a stable, healthy but considerably lower 30 badgers per one *hundred* square kilometres. And camera-trap studies on badger setts here suggest that there are two reasons for this – one is a felid, and the other is a canid.

In Naliboki, lynx visit badger setts year-round, often waiting patiently until cubs emerge from the den, before striking. Impatient lynx, however, will sometimes try and scare badgers in the sett, such that they run out from one of the other entrance holes. Wolves most visit badger setts during the denning period; if just a few badgers are at home, wolves will often predate them, although they rarely bother to decimate a full sett as this takes them too much time. In May, some badgers abandon their setts completely as soon as wolves move in to start denning here. The effect of wolves and lynx upon badgers is thus twofold. Firstly, by killing badger cubs, wolves and lynx both reduce the reproductive rate of the species. In Belarus, the majority of badgers at any one time do not have active litters. Secondly, wolves and lynx continually displace and move badgers on, which further interferes with their life cycle. This renders the badger considerably rarer, and more of a low-density animal, than here in Britain where, in our predator-free and simplified landscape, the badger attains some of the highest densities in the world. However, by reducing both foxes and badgers within a landscape, wolves and lynx open up new outcomes for a whole range of other species.

We have grown up in a culture where cheeky foxes and Badger, immortalised in *The Wind in the Willows*, are, for most of us at least, treasured creatures – and there is perhaps no harm in this. Species like badgers are smart, resourceful, striking and their presence sends a thrill down the spine of any observer. Such sentiments are, however, unlikely to be shared by a nesting woodcock or young hedgehog. Badgers are, in truth, just one nicely striped player in our ecosystem – an ecosystem designed, over millions of years, to be regulated by other, more powerful cornerstone species than badgers; species, which, in turn, free up space for the small and vulnerable to survive.

The complex effects of wolf and lynx may never again be felt in all or even most of our landscape. As our woodlands regrow, it seems likely that the lynx, ghost-like and invisible, shunning open habitats, pasture farmlands and us, would integrate more easily into the British countryside, in those larger wooded areas quiet enough for it to make its home. As for the wolf, it may be decades, or more, before a nation still struggling with beavers, and boar, can bring itself to accept an adaptable, brilliant and far-ranging predator; a predator that will, undoubtedly, take livestock, and in other ways make its powerful presence felt within the landscape. Cultural change, technological change and a far greater acceptance of what wolves do, and the benefits they bring, will not develop in the course of one generation alone.

As this is written, wolves have been watched quite harmlessly scuttling through the streets of German Saxony in daylight, as little feared by most as a fox on the streets of Bristol. Other cultures are already adapting their attitude to wolves – and in place of outright fear and ferocity, there is not only acceptance but also, in some cases, mere indifference, too. It is unlikely, however, that as long as intensive sheep farming persists at scale, that indifference or acceptance will reach every sector of society. The question is whether it must – or whether it should. But other species, like the lynx, leave a far lower social footprint. It seems most likely that their time will come first.

One day, as we grow ever more aware of the need to repair our natural world – a journey that is already decades underway – and the small but loud lobbies fearful of wild animals grow ever quieter, as their own industries diminish or change over time, there will come a time once again where we have the space, mentality and wildness to live beside the only hunters to match us in skill; those that we meticulously hunted out centuries before. When we do, the natural world will grow ever richer, and, as our own acceptance stretches, so too will the sea of trees and layered song that belong to the realm of the Hunter.

CHAPTER NINE

Humans

Golden by dusk, white by night, a barn owl floats across Salisbury Plain, her disc-orbed face turning as she drifts, listening – acute. She can discern the shuffle of a field vole, but the line of sight is far from clear. She works her way along the tarmac path that splits the firing range in two. Here, she has been successful many times. Voles have friends. And when friends decide to meet up, the owl may be in luck. She continues her familiar, linear journey, waiting for that one impatient moment when a vole puts itself on display. She quarters relentlessly. The whole grassland is squeaking – this is a lottery of numbers. At last, a huddled brown form appears on the tarmac path. The owl flutters, pauses, drops like a pillow in free fall. Seizes supper. She has three hungry mouths to feed and heads home; effortlessly cutting the miles on taut, elastic glides. An ancient oak looms ahead.

As the owl approaches, she does not pause to reflect on the extraordinary nature of her nest. Her home might best be described as a tree within a tree. Hard-edged, the pale wood has been hewn by a creature who presumably spent some time up here, hard at work, but has since vanished from sight. The dark, square hole is now lined with three pink and white mottled faces, midnight eyes expectant and demanding. The wooden cave appears to have sprouted from the limb of the oak, yet is much younger than the tree. The owl barely pauses on arrival. Gifting the perfectly preserved vole to her scrum of chicks, she has barely alighted before she takes off again, leaving the owlets to scrap it out within their neat, square home. On closer inspection, the owl's house becomes stranger again. Its wooden flanks incorporate a second door, but that door has been sealed shut. The roof is resinous and the freshly fallen afternoon rain has beaded on the top. And studding this lofty tree cave together are serried ranks of bright, steel screws.

Millennia before, as early settlers arrived in numbers to hunt the wild horses, cattle and deer of Salisbury Plain, barn owls may have been far scarcer even than today. Sixty per cent of Britain is estimated, prior to the Bronze Age, to have been covered in various forms of scrubland, wood pasture and dense-canopy woodland, where the tawny owl and perhaps the eagle owl would have reigned supreme. Our fenlands would often have provided feeding grounds for barn owls, but seldom a safe, cavernous home: on a natural floodplain, giant veteran trees like willows would have been not only uncommon but also hugely contested. The barn owl is a gentle bird. Many others, even jackdaws, can muscle it from its home. But tawny owls, and eagle owls, especially, will kill barn owls outright. So too will that red-eyed woodland fury, the goshawk. Barn owls may always have been uncommon in the onetime natural world here. Now, though they have been endangered by modern farmland practices, there is another keystone species shaping their fortunes, their hunting patterns. A species that now crafts the nesting sites of most barn owls in Britain. Ourselves.

The barn owl nest box is one of the endless inventions that we, the world's dominant ape, have invented to not only mitigate the effects of our occupation here, in Britain, but to actively provide homes for the creatures around us. Our ingenuity in this regard, especially here at home, has been profound. In the UK alone, we can house more than 50 of our regular 200 breeding bird species in nest boxes, and many of our bats. We are the only species on Earth to have consciously created hedgehog dens and newt ponds, false banks for sand martins, fake nests for swallows, scrapes for avocets, gardens for butterflies, green roofs for insects and perhaps, most wonderfully of all, patrols and crossings for the common toad. There is nothing that, if we put our mind to it, we cannot invent to help the natural world – and, in some cases, to better nature in doing so.

There is also no doubt that whilst beavers and whales, bees and boar, free-roaming herbivores and their predators and the wild wonder of regenerating trees can rebuild our shattered land, there is only one truly dominant species on our island – and that is us. We must be the guiding hand, the steward of stewards. Whether with firmer touch on the lands where we live, or farm, or lighter touch in those lands we leave to self-generate, self-govern and grow ever wilder and more self-sustaining over time – we remain in charge. Beavers cannot release themselves, wolves cannot swim the channel, and no species can recover to govern ecosystems, even the great whales, unless we allow them to do so. In this regard, all cornerstone species answer to us. We have become, and remain, the greatest cornerstone species of them all. We can reshape landscapes, and wild lives, on a scale that even the vast elephants of times past could never do. We are, collectively, the most powerful of megafauna ever to have grazed and hunted the Earth.

Until the past thousand years, it might be argued that we have played this role with at least some kind of balance. Whilst the giant animals fell to our need to eat, many of the smaller ones survived. Whilst we felled large areas of woodland (indeed, most British woodlands were gone before Roman times), we did not erase the woodland ecosystem. Whilst we farmed widely, we also created rich grasslands: we reformed the land and reshaped the fortunes of its species, but we did not erase. Whilst we hunted our marshes, and most probably ate some species, such as pelicans, out of existence entirely, we had not mastered the art of draining land. Water, and rivers, were a harvested resource – but not imprisoned, or removed entirely from the landscape, by human activity. From the land, we hunted whales at first in small numbers, harvesting close to our shores. We were a pre-industrial species; one growing in numbers but not, as yet, entirely dominant. But as we grew ever more so, our wish to shape and control the land around us grew.

Quite why Britain's apes exterminated so much life, so early on in our history, and so completely, is a question we may never fully unlock the answer to, but the first and most obvious fact is that we were island colonists. When hunter–gatherers followed the herds across Doggerland into what is now Britain, they had no idea we would soon be cut off from our neighbours – but cut off we were. Whatever wild animals, unable to fly (or make boats), were here, were here to stay and, once removed, would be gone forever. Horses went first, then aurochs and elks. Bears followed, then lynx, beavers, wolves and boar – and others, like the great whales, were eroded away, first around our coasts and then further out to sea. Trees and soil, essential for our own survival, were treated more gently in places – reformed to suit our needs. Grazing animals, too, became reformed to our will. But our island nature increasingly left little room for wilder contenders to the throne.

In spite of this, we became systematic, methodical and determined in our eradication of some cornerstone species very early on – though rarely for one, cohesive set of reasons. Beavers were hunted for their fur and castoreum, whereas wolves were hunted as competitors (and for fur) with a ferocity that outmatched our European neighbours.

By the Middle Ages, our desire to tame the natural world around us grew in intensity and purpose. Those creatures, such as wildcats and eagles, which had until now shared the food-rich environment we had stewarded, at times quite well, would be hunted down under Tudor vermin laws. In each parish, bounties placed on our wild creatures proved devastatingly effective in their eradication, often, like the chough, included within the Vermin Lists and eradicated from huge swathes of our coastline entirely. And so, our role, even 600 years ago, became ever more geared towards outcompeting all other animals and destroying the remaining wilderness around us. As we began to industrialise, the Great Fens would be drained and our conquest of the land-at-scale, beginning with our deforestation in the Bronze Age, grew almost complete, as we mastered at last those areas of land once dominated by the power of flood.

Even centuries later, we would still prove useful stewards of a tamer version of the natural world. Prior to 1800, the countryside still held a teeming aspect that none alive can now remember. Hay meadows, orchards, small fallows and woods, unchecked streams and brooks and lightly grazed farmlands still recalled many aspects of habitats created by the original stewards. Osier beds became accidental nods to the coppicing actions of a now forgotten rodent, thorn-studded farmlands recalled a time where cattle shaped large tracts of land, before hedgerows, confinement and fields. As we heavily hunted the deer in our woodlands, we would have effected change in woodland flora, recalling, to some extent, the long-forgotten influence of lynx and wolves. And whole families of fauna, from bats to insects, were yet to feel our wrath. Cockchafers were so abundant that upon dying *en masse*, they could clog the wheels of riverside mills. Walls of white butterflies from France moved in visible waves across the Channel each summer. The air would have been black with swallows and the bees loud. Life itself remained – with us, beside us, and in extraordinary abundance. Abundance of this kind is, in fact, something that most of us alive today have never witnessed in Britain, outside of a few refugia, like our seabird cities, where it still remains: preserved only on the very edges of the land.

By the early nineteenth century, we had become the first truly industrialised people in the world. Everything would change; our role as stewards would transform forever. As the countryside became a means of yield, feeding ever more hungry mouths, the small, coppiced, messy and varied world of our farms would grow ever simpler, ever more homogenised and ever more silent. Pastures, once untilled and anthill-rich, would give way to croplands; rivers would be locked into straight courses by feats of brilliant Victorian engineering; as destructive to the natural world as they were, and are, impressive feats of human ingenuity. Railroads and canals allowed resources to travel swiftly across our lands and overseas. And the air itself transformed, as factories blackened our lungs and the sky. The Industrial Ape was born – and our role as a beneficial steward, even some of the time, was coming ever closer to an end. Whales were hunted as an industry; entire species, as large as the blue whale, consigned to history in our waters.

By the early twentieth century, we had for the first time the ability to create total order in the world around us; order that often made way for just one species' progress and stability: ourselves. Chemicals, the perfected outcome of a century's industrialised progress, would allow us to reset the soil each year, and eradicate the tiny competitors, like the cockchafers, that we had overlooked for so long or, indeed, had been unable to remove even had we wished. The farm became, or had already become, an industrial unit. Old shepherding ways, whereby animals were moved constantly through landscapes, effecting variety in the farmed landscape, gave way to ever larger, static and industrial herds, or flocks, who mowed everything to the ground. Sameness replaced variety in the tilled landscape as much as the herded one. And over time, the need for urban conformity, as well, would turn our ivy-clad villages into ever tidier affairs. Everywhere, as sameness grew, our capacity to allow life, or even ignore it, was vanishing at an alarming rate.

By the late 1950s and early 1960s came a critical moment in human awareness. We began to notice, at last, the growing silence around us. By this stage, the environmental movement was nascent – but the industrial movement to rid our lands of all but food, order and us was rampant. It would take further decades before, at last, the Ape Aware would emerge.

In the past 50 years, in Britain more than in most European or indeed global countries, our divorce from the natural world, and our role as its steward, has continued apace. Disastrous policies paid farmers to cleanse those last habitats, such as orchards, hay meadows or fallows, which had acted as refugia for centuries. Commercial forestry blackened the land, but no agency funded the recovery of our relict ancient woods. Wading birds, prime indicators of our relationship with the soil, vanished as farms were drained, wet corners made dry, soils compacted and the insects within them hammered remorselessly. Marginal species, whether hedgehogs or whinchats, became history as margins, too, became a thing of the past. The countryside had become a factory at last, and each decade, that factory was perfected. The effects are still being seen every year. Cuckoos, once common in farmland, have become starved refugees, clinging on in those landscapes where the small things, like hairy caterpillars of large moths, survive in abundance.

If our relationship with the natural world is, for the most part, fundamentally broken, then we have at our disposal, albeit not always in one place, or one person's mind, all of the tools to fix it. We are an industrious and inventive people, and whether in the farmed environment, the wilderness, the urban world or the infinite blurred margins in between, we have all the tools needed to become the greatest steward of them all. The greatest tool, however, is human ingenuity. Right now, as the great Sir David Attenborough once said, 'We are the cleverest species on our planet. Now, it is time to be wise.'

Whilst elephants, beavers, boar, cattle, bees and many other species that did, do or could again govern areas of our land, are possessed of sophistication and intelligence, no species is quite so capable as ourselves of using intelligence to good effect. We alone can design cities of a complexity to beat termites, or create musical scores so complex yet absolute that symphonies such as Mozart's can endure for hundreds of years. We alone can translocate other species around the world, design large-scale environments to the finesse of a millimetre, invent global travel, standardise global language and communicate remotely. Restoring the natural world around us should, on paper, be far easier than the vast majority of our species' other achievements.

To date, the greatest problem worldwide has been that, as we have grown, the natural world has shrunk. As such a new species, we have failed to realise that our own survival is dependent not upon ourselves, but upon the natural world around us. That we are not an observer of nature but yet another species – one gifted, perhaps not for longer than a few thousand years more, to govern and control the planet and its resources. We are a species within nature: the cornerstone of cornerstones. As we begin to realise this role, a whole range of possibilities emerge.

In this book so far, we've examined how much certain animals can do to help us repair the natural world. Whilst every species in Britain has its place, the importance of some, their role, significance and effect, is undeniably greater than that of others. The whale creates worlds: the puffin lives within them. The beaver creates ponds: the otter, fishing in those ponds, does not. The more cornerstone species we have, the richer our world will become. But to give them space, and concede them power, brings us to that most fundamental of human flaws – ego.

Having tamed the natural world so early on in Britain, we have, in the past few decades, fallen well behind the efforts of many neighbouring countries when it comes to protecting and restoring it. Creating as much space for natural self-governance does not come easily to a country, or its people, so accustomed to near-total control of the landscape. But the more room we allow for other stewards to flourish, the more they will reward us in turn.

Fishermen and the fishing industry cannot create fish shoals from scratch, to be responsibly harvested by ourselves, seabirds and other fish – but whales can. It is in all our interests that whales proliferate, singing once more around our coasts, visiting our harbours and enriching our waters. It is in the interests of small upland farmers, especially in the ever drier summers to come, to keep water on the land at all costs. If the heritage of Lakeland farmers is to survive, they would be wise to call on the beaver. Given just 10m of space either side of a small river, beavers will give our upland farmers the aquifers needed to defy climate change – and carry on farming against all the odds. It is within the interests of arable farmers for bees to proliferate within our fields; only agribusiness, and short-term gain, now shout for the chemicals that destroy wild bee populations and dismember their minds. Indeed, almost all of the wild forces in this book, except, it must be said, the wolf, enrich most communities in which they are allowed to flourish and survive. And even wolves, through protecting new woodlands from deer browsing pressure, may in the centuries to come become allies of our foresters as, beaten by technology, wolf packs can no longer reach our cities or our farms. Others, like soil and trees, are simply essential for our own survival.

In other countries, especially in eastern and central Europe, some cornerstone species, from beavers to wolves, have simply crept back in. Over time, most have been accepted, and only a few, predominantly the wolf, remain a polarising force. Reactions to wild boar in France, or beavers in most of Europe, even in countries as crowded as the Netherlands, never reach the levels of fear and distrust we see amongst certain countryside lobby groups here in the UK. Our fear of other species remains greater than in most other parts of the world, and at times even a water-retaining rodent can become the subject of sustained and intense persecution. We have a lot of ego to let go, if we are to accept that other animals, beside ourselves, can shape and enhance our shared natural world.

One problem in this regard is that those deeply ingrained within our present countryside, whether farmers or gamekeepers, are invested entirely in a world that they know. For centuries, the landscape around them has been tamed, the green fields incrementally more silent but the general processes of managing the countryside for maximum yield – unchanged. The roles played by many land managers, farmers or keepers are time-intensive, and as such, do not allow for the travel needed to see that in many other parts of the world, from Scandinavia to Germany and the Netherlands, activities such as farming are simply integral parts of a thriving, wilder, wider environment, not a force to which all other wild animals must bow – or bow out.

Another part of regaining our role as stewards is to see ourselves as the Master Delegator of the animal kingdom. Why intensively coppice trees if a beaver can do it for us? Why carefully plant a meadow if a small harem of wild horses or herd of cattle can, over time, achieve the same effect, over a greater area of land, free of charge, using mechanisms for doing so that have evolved over millions of years? We must not feel so threatened, perhaps, by those who help us. Indeed, it is continually surprising how many British conservationists distrust or dislike wild boar, given that pond-digging, soil rotavation and fruit-tree dispersal are all activities that we, as human conservationists or farmers, cherish, respect and have come to perfect ourselves over time. Perhaps it is time to reframe a part of our role as the master steward. Rather than stepping out entirely, we are simply delegating; assigning to other species the job description to which they are best suited. Perhaps this, too, allows a little more dignity for us – and a little more room for our egos.

But becoming better stewards does not simply mean allowing cornerstone species more of a role in our environment – we are eminently able to create environments ourselves, as surely as the great apes of the Congo shape its clearings, plant its forests and aid its species. We are not merely a steward of others – we can create habitats in our own right. And if we can begin to create the right ones, in the right places, then we will once again honour our role as the greatest steward of them all.

Let's start small – with the very ground below our feet, with each square metre put before us. Let's look down to the soil as a magical entity – that in itself is how every farmer once thought, in the age before fertiliser, before nitrates, before rigging the odds, before planting crops like maize that would, over time, leach the soil itself from the land. By looking to the soil, we appreciate there is a world beyond our feet. If each square metre becomes special, that engenders, over time, a very different mindset about how we use, or misuse, larger areas of our land – and so we must look to the soil as we seek to restore our relationship with the other species around us.

We do not owe our lives to crops, our settlements to stable food or our futures to plants; all of these depend upon the entity that has allowed us to farm, grow, develop and rule – the soil beneath our feet. In the past few hundred years, that soil has changed immeasurably, in its quality, composition – even in how much of it remains within our fields. We have forgotten how the soil should feed not only us, but also the wild creatures around us; what soil should sound like, look like and the dizzying array of wildlife it once sustained.

After centuries of de-wilding the land, British soils, until the invention of agricultural industry, appear to have stood the test of time better than many of the other environments that we had either degraded or removed. The soils of 100 years ago were still home to an extraordinary diversity – and abundance – of beetles. Cockchafers blackened the air; in 1911, more than 20 million individuals were collected in just 18km² of forest. We know that our pasture farmlands, until even the last four or so decades, reverberated to the tingling electric calls of lapwings, tail-drumming of snipe and the haunting melody of curlews; species intrinsically tied to the damp soil, and the rich, accessible pickings of invertebrates found beneath.

But, as early as the agrarian revolution, things began to change. Perhaps most significantly, for Victorian times, we began to convert huge areas of low-intensity pasture to cropland. And as we did, we began to disrupt something we would not understand for almost a century to come; the sheer complexity of fungal networks below the soil. This came to pass as we began to till – and to plough, with new intensity and new technology.

From the advent of the Rotherham Plough, invented in the late 1600s, we had been ploughing the soil, but we had, no pun intended, barely scratched the surface. Early forms of ploughing may have turned over the soil, but did not fundamentally rip it apart. Also, until the widespread growth of arable farming in the mid-nineteenth century, huge areas of pasture land, used for extensive and commons grazing, may never have been ploughed at all, containing, like the traditional orchards or pastures of today, or areas like the New Forest, an undisturbed subterranean world of fungi, beetles, moles and larvae, all working away busily below our feet. But over time, the plough would change.

By the mid 1800s, horse-drawn ploughs would become gradually redundant, as steam-powered ploughs were more effective. Then, by the 1920s, our relationship with the soil would change forever as the Ferguson tractor was born. We could now cut ever deeper, and faster, through ever larger areas of soil. Over the past 100 years, especially the past four decades, deep ploughing has begun to endanger the very soil itself. Mycorrhizal fungal networks, circuit boards of immense complexity, which even allow plants and trees to communicate below the ground, are ripped from the ground if soil is deep-ploughed. This can dramatically affect the health of trees such as oaks, whose deep-seated root systems reach out into farmland fields. Fragile invertebrate communities, too, are repeatedly reset from scratch, whilst invaluable networks of rodent burrows, some of which, like vole holes, become the home of bumblebees, are dismembered time and again. Deep-ploughing, creating oblivion of the soil biome itself, fosters desert more effectively than most other methods of farming. But this is not the only war we have waged against our soils and the myriad of creatures within. Indeed, it is only by looking to the farmland of other countries, less intensively managed than our own, that we see what we have lost.

If you drive along a roadside in eastern Poland, Hungary, Romania or many countries that were formerly in the Eastern Bloc, and whose agricultural practises mimic many of ours from more than 100 years ago, you may notice something unusual to the British eye. Not only will those roadside verges be invariably richer than our own, alive with flowers, butterflies and bees, but they will be shorter, too, with large areas of rich, open soil. The soil itself is replete and intact. Invasive chemicals – nitrates and phosphates – have yet to soak into every inch of the landscape through which you drive. Over-trampling, which can compact soil and render it useless to foraging ground-feeding birds, is rarely an issue when smaller herds of animals are herded through the landscape, and rarely kept, in large numbers, in one place for long. Anthills and beetles abound – pull over your car beside the road and you will see industrious dung beetles, or many other kinds, crossing the road, all bent on little journeys of their own. Watercourses often sparkle with clarity; in the absence of phosphates and nitrates, a fine array of plants interweave in an ecosystem unbroken by the rampant, dominant growth of just a few species. In much of eastern Europe, the soil itself is still intact – and alive.

What eastern European farmland reminds us of is that in Britain, if you once left soil alone, in a time before fertilisers washed through the land, and put just a few animals to graze, or rotated your crops every year, affording valuable protection to the earth itself, the soil would become an organism of fantastic richness and complexity. Indeed, the German conservationist Werner Kunz, in his book *Species Conservation in Managed Habitats: The Myth of a Pristine Nature*, explores how eastern Europe's farmland now contains one of the most precious and vanishing habitats in Europe: *soil*. He makes the point that whereas nitrate fertilisers dominate western Europe – promoting the growth of rank grasses – the light-touch farmland of eastern Europe keeps the soil itself alive. Earth is nature's starting culture, in which anthills, dung beetles and many other invertebrates come to thrive. In Britain, however, since the 1930s, fertiliser has become the signature word of our farming. This has affected our landscape in a way that now shocks nobody alive.

If you look more carefully next time you drive through the southern English countryside, the rampant growth of vegetation is striking. Our roadsides grow dense with rye grasses – the agricultural triffids of our time. Our communal parks, which in eastern Europe are often filled with bare, muddy areas and nest-building swallows, are lush lawns thick with just a few species of dominant rye grass. We completely take this for granted, but the pre-fertiliser state of the British countryside was not like this at all.

In an organic landscape, untouched by nitrates or phosphates, which now permeate our water supplies and rain from our shared skies, a huge variety of flowering plants, such as yellow rattle, contest the growth of grasses. Fertilisers have, however, transformed the British landscape – our farmland fields, fallows, watercourses and even our wilder areas – incrementally, but beyond recognition to any farmer who lived a century before.

Both nitrate and phosphate fertilisers act, in effect, to rig the productivity of the soil. They create pulses of sudden growth by providing fast-growing species (such as many crops) with the nutrients needed to thrive; often in the same place, year after year. Prior to the widespread use of fertilisers, crop rotation was required if arable farming was to stay viable. Under this older system, each crop planted would deplete certain nutrients from the soil. The next year, that field would be rested, allowing the nutrients to recover, or planted with a different crop – one which put the missing nutrients back in. Rotation is still practised more widely in areas of eastern Europe, and elsewhere, where the large-scale use of fertilisers has not been adopted. The result is a soil that is naturally rejuvenated, rather than artificially injected with the arable equivalent of growth hormones. Like most chemicals designed to promote monocultural crop growth, both nitrate and phosphate fertilisers have profound abilities to create sameness within both our soils and the life that grows within them.

From the centre of our fields to our roadside verges, what was once a chaotic jumble of poppies, cornflowers and many other arable 'weeds' has been removed, not only by targeted herbicides in recent decades, but also by decades of our rigging the soil itself. Nitrates and phosphates favour and create intensive growth of the few, over the slower, more-rambling growth of the many. As a result, dominant grasses now choke many of our landscapes; our watercourses, field margins and fallows, in place of a more varied fauna of plants. This sameness of vegetation also greatly reduces the heterogeneity of breeding sites for birds, and the nectar sources for a whole array of invertebrates. But fertilisers also achieve something else. In triggering sudden pulses of intensive growth, they effectively create grasslands where the soil is shaded, and its invertebrate treasures hidden from view, and this apparently subtle effect of fertiliser use has had profound changes upon our ecosystem.

Back in the time of John Clare and his local breeding wrynecks, and until the start of the twentieth century, farmland Britain was home to a wide diversity, and great abundance of species whose existence was predicated upon varied, open soils. Red-backed shrikes, beetle specialists, are visual hunters; targeting ground-dwelling dung beetles visible to them in pasture on the ground. Yellow wagtails may nest in dense grass cover, but they probe in open soils for food. Nightjars, still farmland birds in the early twentieth century, hawk for longhorn beetles over open soils, and on the ground. Starlings are heavily dependent upon leatherjackets.

Our fauna contains a wide array of species, from small invertebrates up the trophic cascade to their predators, who require short, diverse swards in order to survive. Whilst grazing herbivores can provide these conditions, fertilisers rapidly take them away, promoting a sudden jungle of overly dense vegetation, which immediately prevents the feeding of many other species. Anthills, for example, struggle to form in nitrate-sodden fields, and those that do are rarely discernible to species that feed upon them. By allowing a few dominant species of cultivated grassland to feast, fertilisers like nitrates and phosphates effectively prevent the functioning of an entire ecosystem of soil-feeders, evolved over millions of years in ecosystems where grasses and flowers grow surely – but slowly.

Today, you have only to drive through the eerily deep-green fields that cover much of lowland England to see the ultimate result. Here, at a landscape scale, nitrates are in charge of our soils – and the single-species green we see is not life, but its absence. By contrast, a diverse grassland, seen from a distance, or perhaps discerned amid a sea of fertilised farmland, is a more yellowed, chaotic affair; a place whose subtly shifting colours bely a wide array of plants and grasses, still intact, and competing, as they should be. But as more and more of our fields turn a deep pure green, that is a sign not of verdure but the slow extinction of the soil, as just a few minerals now dictate the few plants, crops or grasses that can thrive. Today, nitrates are so prevalent that they are carried in our clouds, as rain, falling, even in areas like the New Forest, where arable farming has never taken place. Fertilisers have changed our soils forever, even beyond those areas where they are applied.

Just as we no longer trip over anthills by the dozen, or pause to watch a small army of dung beetles exit a local cattle field, we are now beginning to forget the vast hatches of craneflies in our farmland fields; begotten of leatherjacket larvae below the ground – or, indeed, the hordes of nesting starlings that feed on those leatherjackets, too. Starlings, like wrynecks, are creatures of the soil – and they, too, are becoming increasingly imperilled within our agricultural system. And this comes down not only to the dense jungle of grasses that hide their feeding areas, or the compacted soils that inhibit their feeding, but also to the use of chemicals so effective that they are transforming our countryside entirely.

Of the many herbicides that have effectively weeded species such as turtle doves from Britain, alongside the vast majority of our grey partridges and other species dependent upon arable weeds, glyphosate stands out as the extinction king. In the past few decades, it has transformed the ease with which we can weed our soils of life, but it also sinks far deeper than we may care to think. Above the soil, glyphosate is startlingly effective at the job that intensive farming requires it to do; removing 'weed' species prior to crop planting, and therefore allowing the target crop to flourish unhindered by cornflowers, poppies or indeed almost any other plant. But glyphosate doesn't only starve our wild species on the surface by removing food-plants and their associated insects – indeed, much of its devastation is far harder to see, and happens below the soil. Here, glyphosate reduces the activity and reproduction of earthworms. It is well known that earthworms act as the miniature ecosystem engineers of our soils; cycling nutrients and moving organic residues through the soil, and greatly enhancing soil decomposition. Now, even they are not beyond our reach.

If the war against our soils has been well documented when it comes to herbicides, our war against dung – and the rich world that dung once created – is an altogether stranger and less familiar story. In short, once upon a time, as animals moved, they passed rich dung into the soil. This happened for millions of years, and it happened well into the twentieth century, when cattle and horses were still followed by hosts of swallows, and red-backed shrikes still gleaned beetles from around cowpats. Then, something happened: we began to medicate our animals, and developed worming chemicals such as avermectins. This has transformed our pastures and grazing areas, and the fortunes of those species that depend upon them, not only in Britain but also across much of the developed world.

From the beetle-feeding little owl to cuckoos and nightjars, a whole range of avian invertebrate hunters, not to mention terrestrial mammalian snufflers like the hedgehog, depend upon animals like dung beetles to survive. Prior to the 1970s, analysis of the stomach contents of curlew – then a common bird and today endangered – showed that dung beetles formed a particularly vital food source; perhaps little surprise for an open-grassland and pasture species adapted, for millions of years, to feed beside and around grazing animals and the habitats they create.

Organic pasture systems, like those seen in some parts of our uplands, such as the cuckoo-rich, extensively grazed pastures of some Dartmoor valleys, still retain the conditions for dung, and beetles, to thrive. So too do some areas of northern England, where free-roaming cattle vector dung – and the varied life that dung created – across landscapes rich in the curlews and lapwings that thrive around beetle-filled pastures. But now, all too often, our pastures have fallen silent – devoid of little owls, curlews or lapwings – and that is in part because the very soil, and the very dung within that soil, has been chemically silenced. Across Britain, much cattle and horse dung is now bereft of life. The basis of a food chain, an ancient trophic cascade that once involved elephants and aurochs, and continued until recent times in our farmed environment, has been ripped out. Dung is often now mere dirt.

By worming our cattle, we ensure that in place of grass-rich compost passing through the animal and into the soil, where armies of dung beetles await (as can still be seen in the bird-loud wood pastures of Spain), instead, toxic chemicals are passed, in great quantities, into the soil. Dung beetle larvae, and many other beetle species heavily dependent upon dung, founder or die when they come into contact with the chemicals, passed by livestock, into the soil. Incrementally, over time, this apparently unassuming process can lead to devastation, wrought silently, across entire landscapes. One of the most extreme, documented examples of this has been seen in Australia. Here, vast herds of medicated cattle were leaving behind huge quantities of medicated dung. With scarcely any dung beetles left in the ecosystem to break the dung down, more beetles had to be reintroduced in order to break down the dung and sequester it back into the soil. Dung beetles are a vital component of the way our soils work. Not only do they provide food for a host of Britain's vanishing species, they also move dung, and its power to create and enrich floral life, deep into the soil. Without dung beetles, the very means by which free-roaming animals vector nutrients across the landscape is removed. The ancient Egyptians rendered the dung beetle a god – and we ignore its importance at our peril.

Dosed with chemicals, with the pharmaceutical drugs passed in cattle dung, and filled with waves of phosphates and nitrates, Britain's soil is not only washing away into the sea, and from farmland fields into our rivers – but also being

sterilised within an inch of its life. But by looking to the
rich life it once held, to the more-traditional farmland
systems that endure in parts of Europe, and to the future we
want for our own crops, our food, and ourselves – the soil
too can be rewilded and restored.

Within just a few years of turning cattle back to grass,
and removing the chemicals used to worm them, soils have
begun to regain their dung beetles. Grazing those cattle
extensively, across soils unsprayed and uncropped, has seen a
return of anthills. As arable farmers have turned away from
deep-ploughing the soil and towards a policy of 'no-till', so
the mycorrhizal fungi that communicate between the roots
of plants and trees, promoting their health and allowing
them to network underground, have returned. And as
rotating crops has become, once again, more normal
practice on many British regenerative farms, so nutrients
depleted in the previous year have been returned by the
next crop planted, and the soil held in place for future
generations to farm it. Given any chance, the soil remains
the most robust of all wild forces in Britain. Now that most
of the great whales and all of the wolves have gone, it will
take great effort, expertise and cost to put them back. The
soil, however, is a far more forgiving entity – if we allow it
to recover, in time.

Soil is the substrate upon which many other keystone
species depend. Grazing animals pass dung into it, fertilising
it, vector seeds around a landscape – which fall into it – and
create wallows, which, in flooding, become ponds for new
generations of aquatic invertebrates and amphibians. Boar,
in rotavating worm-rich soils in a growing number of our
woodlands, can reset and transform soil into a chaos of new
floral life – but only if drawn to the rich contents within.
New plants and pioneering trees will colonise soil without
human help, but their diversity, complexity and balance will
founder if the soil has been rigged with nitrates or
phosphates. Beavers, in flooding areas, and leaving damper

soils in their wake, create the conditions for willows, aspen and rarer trees, like black poplar, to colonise – but only if the soil itself retains the ability to feed – and anchor – these species. Without healthy unfertilised soils, the zealous action of even a billion healthy honeybees will come to nothing, as the rich array of nectar-bearing plants they require are replaced by ever poorer monocultures.

Soil is self-evidently vital to the natural world and human life. As we begin to safeguard it once more, we will begin to rekindle the conditions to rebuild the world around us. Then, both we, and our fellow farmers and engineers, can once again get to work – creating a richer world for us to feed and thrive within.

In the past few years, we have seen many inspiring movements towards restoring our soil. Regenerative farming, a phenomenon barely heard of at the turn of the century here in Britain, is now a growing force in the countryside. Many farmers, who do not wish to sacrifice yield or turn back the clock and resign the land to nature, are nonetheless moving towards preserving the soil. 'No-till' farming, with protective cover crops, free from inputs such as fertilisers, has become gradually more commonplace, and the soil itself, sustained and better protected, has been able to better nourish a wider range of crops. Healing the soil takes time, but Gabe Brown, one of the leading lights in America's regenerative farming movement, believes that it comes in five key stages. First, we must protect our topsoils at all costs, avoiding tilling unless we must. Bare soils must be protected, and the armour of leaf litter that covers them given shelter. We must vary what we grow, and not leach the soil of its integrity by planting the same, short-term crops in the same place, over and again. And we must avoid intensively grazing just single species of livestock, which trample and compact the soil. This, in turn, leads us to the question of how we might better deploy grazing animals in the future – animals, in the case of the cattle and horses, that once stewarded a far healthier landscape than what we have in Britain today.

Whilst some of the species in this book, such as whales or wolves, or even beavers and boar, are still unfamiliar to much of the British public, cattle and horses are familiar to us all – but the manner in which we, their guardians, have used them, has transformed in the past few hundred years. Since the time of enclosure, animals that once roamed shared landscapes have become confined in fields, and since the 1970s, the number of cattle in those fields has risen exponentially. Penned sheep effect an even greater degree of landscape sterility, razing huge tracts of our countryside to biologically unprofitable lawn. But we need not always graze animals this way.

Grazing animals once roamed continuously through their range, generally in social groups. They would have lingered longer, in greatest numbers, where forage, such as lush coastal saltmarsh or lowland meadowlands, were most to their liking. And on our larger farms, not only in our rewilded areas, we might replicate some of this. By breaking down field boundaries within farms, or communally grazing animals across small networks of upland farms, we can give more choice back to our grazing cattle; instead of confronting them with single fields of grass, by moving them through the farmland landscape we can effect a different set of outcomes. It is fascinating how, given any choice, many older breeds of cattle will browse hedgerows and trees within their fields; woodland, as much as grassland, is their preferred choice of forage. Even now, their aurochs' instincts remain intact. By moving cattle continuously, we allow them to replicate some of their ancestral, free-roaming actions – instead of mowing habitats to the ground, they allow a future grassland to take hold after they have moved on. Of course, such decision-making by us requires farms with the space to do so. In many parts of eastern Europe, and indeed the oldest farmed parts of the UK, a field just abandoned by cattle, or horses, can, within a few weeks, transform into a rich grassland in its own right, provided the herd sizes have not been so large, or concentrated, as to raze plant-life to the ground. On our smaller farms, fences or dry-stone walls might not always be seen as a hindrance to replicating some of the actions of free-roaming animals. Indeed, if continuously moved, and kept in one place for just days at a time, cattle can still participate in 'gardening' a field without mowing it to the ground. It is indeed in these regrowing fields that lapwings and curlews, until recently so at home in the farmed environment, can still find food – and safe haven.

Free-roaming pigs, too, if continuously moved through the farmed landscape, not only have happier lives but can also effect some of the actions of wild boar in our wilder landscapes; disrupting the soil and creating, after they have moved on, rich and varied grasslands in their wake, freshly colonised by flowers. And by exercising our own intelligence in how we move animals, and as our farmers begin to understand more and more about how to replicate wild processes, both small farms and large-scale producers can contribute, once again, to protecting our soils, and putting grazing animals to better use in creating a diverse environment. This is not ecosystem restoration, but it is a viable solution for farmed, productive land – a solution that, by not erasing the soil, plant-life and dependent insect life in farmland, will once again allow the small things to thrive.

That said, we, as the keystone species, have become all too accepting of a country that is, most often, denuded and overgrazed; a country where grazing animals have, through our own mismanagement, run riot. If we are to reprise our role as stewards, we will need fewer, healthier, wilder animals. We will need to stop worming our cattle and rely instead on their feeding naturally, and moving, so as to reduce the risk of parasite infections from their own dung. We will need to reduce our reliance on large flocks of sheep, which denude huge areas of upland Wales, and western England, for few jobs and just a small percentage of our diet. We will need to eat less meat, and better meat, less often – fuelling the conditions in which our farmers can survive, by grazing fewer animals, less intensively, across larger areas of land. And herein lies another challenge for us all. We have become, over time, an intricately complex species. Decisions made in supermarkets percolate down into the denuded countryside in which we live, and the damaging policies we pay many of our farmers to implement on our behalf. Even finding where responsibility lies in such a situation is difficult and complex. The more we demand from the land, and the more factory farming we call for, the more those acting on our behalf will overgraze the land. The less we ask for, the more discerning we become, the more space, money and time we allow for regenerative farming to become, over time, the norm. This, in turn, leads us towards how we might better steward our wilder places too.

Trees and scrublands have often, in the UK, been seen to compete with the farmed environment, consisting, often, of enclosed pasture or croplands. But in many European countries, this conflict has already been solved. Silviculture, the practice of grazing animals like cattle and pigs within woodland, is widespread in areas like the hillsides of northern Spain, where wood pastures, replete with vanished British birds, like the wryneck, are a far commoner form of agriculture than they are here in the UK.

By moving cattle and pigs towards a more wooded diet, we not only create healthier meat, but we also farm animals in habitats closer to those their ancestors remember: woodland and scrubland. Over time, we may, in some areas, be able to make the field structure obsolete once again, creating farms of far greater landscape variety, where animals graze within a myriad of habitats – all of which are biodiverse and varied, but also feed the animal in question, contributing towards the quality of its meat and the gut biome of the grazing animals in question. Indeed, it is most readily apparent on visiting the more biodiverse farmlands of traditionally farmed Europe, that the integration of farmland animals into a wilder landscape is the greatest contrast with farming here in the UK.

With ingenuity, and open-mindedness, we could begin, like the Norwegians, to move our animals into woodlands, and scrublands. And then, over time, to plant more woodlands, and scrublands, across our farmlands too. In this regard, it is often laziness that prevents such readily achievable outcomes from entering the mainstream. Mechanisation has lost us our connection with the land. Herding animals through habitats takes more time, and more labour – but already, that tide is turning.

When it comes to producing our cereals, and other crops, we might also make more space for scrubland and trees within the farmed environment. The vast tractors and combines we now use predicate a landscape free from nuance, one where scrubland islands, fallows, linear rows of riverside trees and hedgerows must all bow out before the mass ploughing operation. But it need not be this way. In the past few decades, we have seen extraordinary miniaturisation of technology, from computers to phones and cameras. Might we not, if we put our minds to it, develop new forms of micro-farming, whereby far smaller fields are productively harvested by far smaller machines – or even large arrays of drones? Our ingenuity often comes to the fore during war, or indeed, pandemics, it would seem, but we are in a constant war to save the natural world around us. With the application of technology, the vast, silent monocultures of modern arable Britain could still cede to diverse, small fields, each growing varied outputs and, in aggregate, supplying the market with just a little less food than at present. Which, given that we waste a third of all our food, would create no net loss to our diets here at home.

The more we apply our intelligence to the soil below our feet, consider what could grow in it, graze it to maximum advantage and understand what varied grasslands, scrublands and woodlands could grow from it, the more varied solutions, for even small areas of land, become apparent. Not all of these solutions are new: indeed, one of the very best mixed land uses – the traditional orchard – has been around for centuries, but quietly ignored in recent times. Orchards provide powerful sharing arrangements between farming and nature. By default, their structure, of fruit trees, protective hedgerow scrub and pasture, provides myriad habitats in one place, and the oldest, rich in deadwood and nesting nooks, can replicate many of the best aspects of wilder woodlands. But orchards can also be eminently practical landscapes within which to work with nature. Apples, pears, cherries, plums, timber, charcoal and cider may be the more obvious products an orchard provides, but its sheltered network of branches shields pasture for cattle, and traditionally, orchards have been ideal environments in which to keep free-roaming pigs, like the Gloucester Old Spot. For one relatively small area of land to yield so many outcomes is perfectly possible, and at scale, the economics and output would add up. Right now, we all pay heavily for homogeny and silence – but we could, over time, begin to pay, as consumers and taxpayers, for local produce, variety and life.

As the only policy-makers in the animal kingdom, we have it within our grasp, all the time, and whenever we want, to put those policies to good use. If just one policy, the Common Agricultural Policy, has in recent times sponsored the homogenised oblivion of the natural world here in Britain, it is just as realistic that new policies, designed to sponsor farming with and within nature, could reverse that trend in a matter of decades alone. We need new policies, which pay generously for native fruit, and for orchards to be put back in. And this is just one of many policies we might create to help us live off the land, yet alongside our fellow species too.

Keeping grounded, we might also devise policies that help the wildlife sharing the urban world around us. Indeed, whilst personal freedoms are important, so too are rules, or human society would, as it has done in the past, descend into chaos. The rules we set in our cities are often quite rigid; they must accommodate, and yet also restrain, the needs and urges of tens or hundreds of thousands of people. Right now, many urban rules exist that wipe out the life around us. Street care and local council regulations enforce the mowing of verges, the incessant tidying of urban hedgerows, the bizarre leaf-blowing of parks, the removal of bramble, nettle and all kinds of other little havens for wild things, and the toxic deployment of glyphosate around us all. But what if we created a new set of rules – ones easy to follow for us, until such a time that we all, once again, attach innate value to the natural world around us? We could, for example, design our lawn parklands to be meadows, not the other way around. We could cordon off scrublands around our rivers and streams, and create allotment corridors in our cities, affording linear safe passage to species from bullfinches to hedgehogs. If we can master the planning of traffic lights and roundabouts, and urban plan to within centimetres, the skills already exist in our minds, and our laws, to achieve the same for the natural world.

At the same time, we might look with closer attention at where we, the Disturbing Ape, can actively benefit wildlife in the human and farmed environment. After all, since the disappearance of the mega-herbivores, the great habitat openers, we are the closest thing left. Even a golf course, grazed by humans for the striking of a ball into a hole, can become a riot of skylarks if its grass is left to grow just a few inches tall. To see ourselves as lost giant herbivores is not only fun, but ecologically useful as well. At the time of writing, farmer James Rebanks, in the Lake District, has begun seeking to replicate the lost effect of 'elephant and rhino' grazing on his upland farm, using the animals still very much available to us today. But there are already other farming systems that remind us how well we can play this role.

Large hay-meadow systems, as can be seen on the Outer Hebrides or islands like Tiree, off the west coast of Scotland, have a unique aspect perhaps never quite seen in nature, but perhaps never bettered by any wild species. Here, hay meadows achieve a density and diversity of flora rarely seen in wild habitats where, even in the old days, nomadic grazing pressure would have thinned out flowers, butterflies and bees. These little havens are our unique creations – we must cherish them, and learn from what we did right, even if those original actions were agricultural accidents. Now, we cultivate and protect hay meadows, redshanks and corncrakes by design. And as a reflective species, when we wish to be, we can now study and replicate our actions. Wherever we have become truly beneficial as a species in our own right, we must upscale those actions, one field at a time.

We have gone so far astray as stewards that even our tolerance of nettles, brambles, scruffy hedgerows and fallows has, in the past century, been lost, so there are many small steps to be taken before some of the big ones can. But as we begin to settle back into our role as decent stewards, the gradual move towards rewilding larger areas, especially our depopulated peatlands, and failing deer or grouse estates – becomes a more viable possibility. With our minds already attuned to playing our own role in the ecosystem with a sense of responsibility, it may, over time, became far easier to accept the skill with which beavers, boar or lynx play their roles, too. And so, from an incremental acceptance of nature, we can begin to super-size the scale and ambition of our actions, in restoring the natural world in all its unfettered diversity, strangeness and abundance.

As we begin to steward more lightly, and relinquish such an iron hold on the landscape, the normality of fallows and their bees, deafening songbirds in our villages, bats hawking moths around our streetlights, and even storks nesting on Sussex chimneypots are likely to become, over time, more and more normal. In central Europe, many villages exist within nature to such a degree that they can be barely perceived from a distance through a mass of trees and bushes. These kinds of small compromises we can readily aspire to, but when we get there, they will begin to add up – not only in terms of a growing abundance of life, but also in terms of a shifting cultural mindset towards the natural world. If we can all accept the small mess left by a colony of house martins, the greater chaos of a local beaver pond – with its frogs, fish and kingfishers – increasingly becomes something that, if not celebrated by all, is feared by only a strange few.

In the past year alone, attitudes in the UK towards one of our most important keystone species – the beaver – have indeed begun to shift this way, for the first time in perhaps almost a thousand years. A modest rodent already quietly doing its thing for decades in Estonia, Germany and even the Netherlands has, finally, been granted governmental safe passage to work its magic, albeit in a constrained fashion, along several British rivers. The beaver is almost back – but it still faces much fear, resentment and, at times, Medieval levels of misinformation. Whilst beavers actively promote the spawning grounds of fish, small lobbies still peddle the myth that salmon cannot jump beaver dams, forgetting, perhaps, that they did just this, or simply swam around them, for more than two million years before beavers were eradicated from Britain. Until recently, other small but voluble groups claimed that beavers ate fish. Yet, over time, like the belief that swallows hibernated at the bottom of ponds, future generations will laugh at our ignorance. Our now ever more detailed knowledge of the natural world is only a recent phenomenon – and as each generation develops and grows, such myths will slowly fade into the past. Provided, of course, that we can keep species like the beaver with us, enterprising, chewing, and healthy, until that time comes.

In this tidiest and neatest of nations, the positive response to beavers seen by most of the public, and most conservationists, represents another step forwards for us as responsible stewards of the landscape; stewards who are now, increasingly, making room for another beside ourselves. Yet it is not only landowners and farmers who fear a loss of control when an animal as powerful in the landscape raises its orange-toothed head into view. Ecologists, too, appear to regard the beaver, at times, with a distrust and fear that is unusual for those who study the natural world. For some time now, the orderly and well-documented world of British conservation has been built around the entirely erroneous idea that only we can create habitats for the other animals around us. Whilst there has been some nod to the ancestral role of cattle through conservation grazing, the idea that beavers and boar can create habitats for bees, butterflies, moths, lichens, fungi and many other orders *without direct human intervention* has been very difficult for the Conservation Ego to accept. Here, once again, we have struggled to let go. Boar, in particular, appear to incense some ecologists with their unpredictable behaviour. Those studying amphibians express particular concern that boar can, and do, eat tadpoles and newts: they do. Boar also, however, create thousands of new habitats for amphibians as their wallows flood, and whilst they will go back and eat some of a new pond's colonists, being opportunistic and not scheduled in their eating habits, they do not eat them all. Such activities, however, are difficult to quantify – like much that is wild in the natural world. And this terrifies a certain breed of ecologist who, in Britain, has evolved, over time, to study the natural world as if it is a zoo, or each butterfly, moth and habitat a static exhibit frozen out of time.

Boar and beavers introduce chaos in place of order, and cyclicity in place of linear results, just as elephants do in the river valleys of Zambia. The difference is that we have become, in their absence, accustomed to a carefully documented and monitored natural world, where each precise conservation action yields a precise result. In turn, the degree of specificity amongst our ecologists is unusual. In the USA, a far more holistic approach to ecosystems is normal; observing not only the aspen, and the beaver, but also the wolf's effect upon the aspen, the beaver and the wapiti. In short, conservation, too, can become myopic – expecting the natural world to confirm to hypotheses, present itself for perfect comparisons and interrogation, or be seen through the eye of a single invertebrate or lichen alone. Yet because the natural world, in its full interlocking chaos, does not function in this way, many ecologists have struggled, and will continue to struggle, with species such as beavers and boar. They introduce something terrifying to the orderly mind – the prospect of unknown outcomes. By and large, the action of ecosystem engineers tends to greatly enhance biodiversity and abundance, but not everywhere, or all the time. Beavers will sometimes flood the habitat of rare moths, boar will trample vegetation, dig things up, and eat a wide range of sometimes threatened species. As native fauna, however, they seldom, if ever, wipe them out, unless that species has already been driven to the brink by a far more destructive force: ourselves.

Rather than pitting animals as powerful, glorious and necessary as beavers against the interests of other single species, of lesser influence in the landscape, the solution surely lies in presenting these ecosystem architects with a playground big enough that all parties can flourish. Given landscapes large enough and rich enough, beavers and boar will finally attain their full influence in the landscape, and those few species who struggle with their presence will, given sufficient scale, be able to move, avoid or adapt to their presence. This brings us in turn to another type of acceptance we, as the dominant steward, need to make here in Britain: the acceptance of *scale* as a vital component of any natural system designed to stand the test of time. Scale is everything.

As we master the ability to steward small areas around us, our gardens and parklands, our regenerating farms and our relationship with the soil, trees and insects around us, it will become ever easier, over time, to confer on other areas – like the vast deer estates of Scotland, or those areas where farming will, inevitably, cease over time – the possibility of reversion to wilderness. Even here, we will not step out of the picture. A wilderness created by us is one that will require our continual involvement. Indeed, even in the most remote parks of southern Africa, elephants are collared, watering holes monitored, rhino patrols sent out, tourism exploited and some commercial resources harvested. Wilderness is not a world without us – it is a place we can create, and then manage in the very lightest of ways. And in those areas of enormous, unpopulated space – even now supporting few and dwindling livelihoods and jobs – the possibilities of scale become exciting indeed.

Areas like the Cairngorms, half the size of Yellowstone National Park, or the collective deer estates of the Highlands, almost twice Yellowstone's size, have been seen for many years as the perfect area for the most ambitious nature restoration projects in Britain. This is not without good reason. There are few roads here. Huge areas of the Highlands, from Altnaharra to Lochinver, and the vast north-western spine of montane Scotland, from Ullapool to Torridon, Lochcarron to Mallaig and Glenfinnan, are, as seen from above, simply empty. Crofts line some of the outer fringes, but the heart of the north-western Highlands comprises deep forestry plantations, which show as the eye is drawn inwards on the map, and, for the vast majority of the area, heather, grass – and deer. This is not a wilderness, nor is it wild – nor is that emptiness good for a species who once lived here in far greater numbers: ourselves. In all, the

Highlands of Scotland, most of all the north-west uplands, all the way to the vast Cairngorms National Park, are sitting, waiting, quietly, for the emptiness to turn into a glorious riot of life.

When it comes to creating a Yellowstone in our own country, there is no precedent for having done so. We slowly eroded our wildlife over millennia – but whilst rapacious in our quest to take it away, albeit often by accident, we have, as a species, been paralytically cautious about bringing it back on purpose. One thing you often hear is that the landscape has changed. In southern England, or the road-crossed moorlands of Yorkshire, this is, to a greater or lesser degree, true. In northern Scotland, forests have been cleared, Highlanders moved in, and moved out, but the space and possibility remains. You can stand in many vistas, from Forsinard to Cape Wrath, and south, for hundreds of kilometres, and in many places not perceive a single human artefact discernible in the landscape. But it takes more than space to bring wilderness back – it will take courage, togetherness and ingenuity.

In France, and especially in northern Spain and Portugal, a great deal of ecological restoration has happened by accident. As farmers have moved to the city, bereft of the generous subsidies that support otherwise unviable farms here in the UK, villages have fallen quiet – and wolves have moved back in. Whilst it might be easy to celebrate such recolonisations from a purely ecological point of view, it would be better, perhaps, to imagine futures where people do not need to give way to the scrub that will swallow former villages over time; where people do not need to cede to animals, nor the other way around. This future is possible, but it will require above all two things. If we are to live beside Yellowstones, with wolves, lynx and elk in the heart of Scotland, we must develop that ingrained respect for sharing the land with other, larger species. But to make that a reality for most people, we must use technology.

On the face of it, ecologically speaking, wolves could be released tomorrow to hunt the Highlands of Scotland. Studies have shown that where the wild prey-base of deer is robust, as it most certainly is in the Highlands, wolves prefer not to venture close to human habitation. At a national level, social or economic, their impact would be negligible for most people, most of the time. Indeed, anyone who has sought wolves in Europe, even where they are relatively common, will have some idea of how extremely shy and elusive these animals are. But, every now and then, lambs, occasionally whole groups of sheep, would be killed. The outcry would be enormous. And an animal demonised for centuries would, almost certainly, be removed once again from the landscape. Furthermore, whilst wolf attacks on people in western Europe, including where the species is present in good numbers, such as Italy, are incredibly rare, the few attacks documented in recent years tend to involve rabid wolves, which lose their fear of people. Whilst, on the face of it, 85 dog-related deaths a year in the UK would prepare us for one wolf attack every year, perhaps less, such a rational approach is highly unlikely to be taken. And after just one attack, wolves would, once again, be removed from the landscape or become, at the least, far less viable as a species capable of living beside us.

This, perhaps, is where technology might, one day, offer a solution. In other words, we might again have wolves, and safely – but if we are, it's going to be expensive. Already, in the USA, wolves have been successfully fitted with collars that not only locate every individual after being fitted (shortly after birth), but also deliver a short sharp shock should they come within distance of a ranch where attacks on livestock may take place. If livestock is not fenced, it would be expensive, but by no means impossible, to collar each livestock animal in turn to achieve the same effect, allowing animals to roam safely, further from farms or fields. In theory, the technology also exists to apply this to towns, houses and thoroughfares – safeguarding both wolves and ourselves. The question is not whether, in this age of space-age technology and phones that can predict and read our shopping habits, whether we can have wolves safely – the question is whether society wishes to afford it. And that, in turn, is where we can look to other countries for how they balance the needs of ourselves against the inherent value of the wild.

Wolves, being charismatic predators, are generally highly lucrative in areas where ecotourism operations can be organised to see them. In some countries, like Canada, some areas with wolves are actually too remote for this to be a reality, but in areas like Spain, or indeed, Scotland, this is unlikely to be an issue. Indeed, the economic revenue generated by wolves would percolate through many sectors of the rural Scottish economy – accommodation, transport, education, guiding and safaris, habitat enhancement, forestry and even deer-stalking, as wolves increasingly weed out unhealthy animals, leaving, over time, ever fitter and larger males for hunters to bag. Most of all, the massive regeneration of trees accelerated by wolf presence, played out across huge tracts of Scotland or, one day, even England and Wales, would protect communities against flooding and potentially save hundreds of millions of pounds every year.

Against this, the impact upon small farms would be significant at a local level, but less so at a national one. In short, therefore, some species that we might currently consider entirely unviable in Britain, are perhaps also those most likely to pay for themselves. The income generated by charismatic species such as wolves, as well as other long-lost herbivores, such as elk, would thus form part of a loop economy, wherein a sizeable proportion of that income was invested back into collaring, fencing and other means necessary to keep all parties safe. This has already happened in many countries around the world.

In order for us to return large areas of the land to nature, albeit with us heavily invested in that model, and communities profiting from it, it is important to remember that we are a species driven by dreams, outcomes and things to look forward to. In this regard, we perhaps never stop being children. Therefore, whilst visions of Scotland returning to trees and scrub is ecologically viable and sensible, restoring landscapes without the enchanting creatures that once roamed them is, to many people, tedious – and unlikely to lead to the tourism investment needed to make such operations worthwhile. In this regard, when it comes to the ecological restoration of large areas in our country, it is important we do not simply create habitats, without the animals that once made them complete. That said, everything comes in stages, and whilst peaceful and well-disciplined coexistence with wolves may be possible in time, another species, like the lynx – shady and shy, silent and furtive, and far less likely to impact widely on farms – is widely regarded as a predator more able to return to our shores. Indeed, areas like the Spey Valley, with its interconnected, deer-rich woodlands of Rothiemurchus, Abernethy and many others all intertwining and inter-joining, in the one part of Britain where wild woodlands and giant capercaillies meet, is, even now, eminently suitable for the return of the lynx.

At present, however, the one keystone species capable of assigning a future to all others is distracted. We have reacted with swiftness to a global pandemic, producing several effective vaccines in just a year, an extraordinary feat. This is because Covid-19 has threatened us, our loved ones, our jobs and our lives. It is understandable that we have reacted more decisively to cut off a pandemic, investing billions, our best minds and best technology, than we have in restoring our natural environment. Yet as more and more droughts and floods wreak ever more devastation on our Atlantic island, paying handsomely for beavers should become imperative for any forward-looking government, as it has been in other countries now for some decades.

As we begin to see the natural world, and the problems it faces, at scale, a range of other policies, far too bold to conceive being implemented at present, may also come to pass. We know now that Europe's catastrophically dwindling invertebrates, whilst small in themselves, require huge areas if they are to survive in the long term. Nature reserves have little chance of holding onto invertebrate abundance and diversity, if those reserves are hostage to a hostile wider landscape. Given that bees, in particular, are essential for many of the crops we grow, protecting them at a national or at least, ultra-regional level, is going to be essential. If we are to protect insects and the soil, we are, sooner or later, going to need chemical bans that extend not across tiny nature reserves but entire counties. Insecticides, herbicides, fungicides, glyphosate, nitrates and phosphates all fundamentally disorder and deplete the basic building blocks of life: plants, including 'weeds', wildflowers, insect abundance and the balanced nature of the very soil itself. The lobbies behind those who sell these are powerful and well-entrenched, but like those who sell crude oil, they are flogging produce that is already approaching history. If we wait for all our farmers to embrace regenerative methods, the natural world may wait too long – longer than it is able. Chemical bans at the 100km level would bring about more rapid change.

Just as no-fishing zones exist to allow continued exploitation of fisheries elsewhere, so chemical bans in some areas would concede the need for some, continued use of chemicals in others. Over time, if more and more subsidies were created for regenerative farming, the need for such bans would grow increasingly obsolete. Furthermore, bans could, like fishing zones, shift over time, allowing, in each area, a large-scale recovery of insect life, making it more robust against any future onslaught. Once again, these outcomes are entirely possible – but only if a national mindset is applied to the protection of the natural world, and that species currently living amongst its relics – us.

Over time, it is likely, however, that technology will create far more uncontested space for the natural world that we have at present – and in two specific ways: the synthesis of meat, and new ways of growing crops. Already, lab-grown meat is on the rise. And rather than rant against it, we might see in this opportunities for all. Right now, many of us pay low prices, through our supermarkets, for what is in effect, already, factory-grown meat. We pay for the wild boar's descendants to be kept in conditions that should never be allowed, and for cattle and sheep, medicated, fed an unnatural diet, and often kept in barns through some of the year, to be driven huge distances in cramped trucks to abattoirs in order to be killed. If we are prepared to settle for meat of such low quality and provenance, over time, it is inevitable that lab-grown meat will eventually eclipse such demand. Our cheap sausages, and bacon, may eventually, and perhaps rightly, be grown in a lab. Pigs, best adapted, in the farmed life, to moving widely through the landscape, may, as farmed animals, become rarer and more prized. Conversely, the quest for high-quality meat such as beef, as something special and worth paying for, is likely not to decrease but *increase*.

Those seeking authenticity are likely to pay far more attention to the farming methods and wider story behind those animals still being farmed on the land. We will pay more, less often, for wilder, better meat, the more so if we know that lab meat can cater to the smaller purse. This, in turn, will benefit those farmers, many of whom exist already, who have removed inputs from their meat and are grazing their animals in greater harmony with the environment. In short, true factory-grown meat may, thankfully, spell the end of factory-farming real animals. And over time, this may liberate huge tracts of the British landscape, either to be extensively farmed, or turned back to nature. Finally, in the farmed world, the balance may swing back to the 'traditional' farm and its values of grass-fed, free-roaming meat, a good life, and a good death. This would be significant indeed.

It may seem hard to imagine now, but hobbies invented mere centuries ago, such as driven grouse shooting, will naturally fizzle out over time. Already, we are seeing Scottish landowners moving away from the intensive management of landscapes for just one quarry – the red grouse – towards more holistic models of land management, whilst each year, more and more grouse moors are being bought up for more progressive purposes, such as mixed use forestry. As it becomes increasingly untenable, socially or politically, to burn land just in order to increase heather shoots, so that red grouse can be farmed in unnatural densities, so more extensive, wilder forms of hunting – like those practised in Scandinavia – are likely to come to the fore. Often, these decisions may not even be imposed on landowners by law, but, increasingly, come from within estates and communities. Such a decision would free up huge areas of our uplands for natural regeneration, and the restoration of lost natural treasures. Hunting per se has never been the problem in our uplands; indeed, it has protected huge areas against development – an impressive feat in a relatively small country. However, the time of canned grouse-hunting is nearing its natural end.

Inventiveness, a human quality used surprisingly rarely when it comes to the British landscape, will most certainly be required to fill the void left when unsustainable sheep farming, factory dairy or intensive grouse rearing eventually fade as core land uses in our country, as slowly but surely, they will. Many raised in the UK, and who have remained tied to areas where they were born, struggle to imagine what could be there instead. Indeed, we have so few wild or

near-original landscapes left, it is often said that if we don't
have sheep, we must have forestry, or, if we don't have
forestry, we must have moorland. In truth, no such binary
choices exist. We can create new, mosaic landscapes, where
native trees are harvested, cattle are grazed extensively and a
proportion of them culled, wild animals can live freely,
under legal protection, and ecotourism, hunting, walking
and many other pursuits can flourish side by side. In many
ways, this description could be applied to most of the
national parks in Europe. It is not a fantasy – it is already, in
most places, happening.

And as we go onwards with more and more courage,
restoring ever larger tracts of land, and larger species, sealing
off large areas of countryside for the recovery of insects, it is
very important to remember one thing. This is not altruism
at work – because the greatest beneficiary will be us. It is not
only about helping other species to survive. It is also about
helping ourselves. When we can once again harvest the soil
and hand it down to the next thousand generations, enjoy
natural pollination and self-regulating insect pest control
through the hordes of insectivorous birds on our farms, and
enjoy meat from both the sustainably farmed and wild
environment – our own health, physical and mental, will, as a
country, be dramatically improved. When beaver populations
attain the density needed to prevent villages from flooding –
as they can, and they will – the rants of a few that a day's
angling has been spoiled will either be ridiculed or, most
probably, ignored. In truth, society will be the ultimate
winner if our other cornerstone stewards return.

Looking outwards from our own island home, we might
turn our attention to the oceans, too, and apply vision and
imagination to how we reconstruct them, and protect and
enhance the life within. The return of great whales is
already happening – but it is happening very slowly.
Perhaps too slowly. Whilst the more and more regular
appearance of humpbacks, fin whales and magnificent pods

of orca in our waters is a cause for celebration, we have all forgotten the veritable whale armies that spouted off our coasts just three centuries ago. We may, over time, invest technology, thought and sensitivity into how we might accelerate their return, too.

In this regard, there is little doubt that the single most powerful ecosystem engineer missing in all of Britain, is one that may now never return by itself – the grey whale. Immense, coast-dwelling and capable of utterly transforming the marine environment, grey whales, being coastal, are likely to have far more dramatic effects upon our seabirds, through the proliferation of plankton and sand eels, and upon the trophic cascade as a whole, than any other single species of marine mammal. Grey whales, however, have been lost entirely from the Atlantic – and those in the Pacific, like so many animals born with migratory instincts, are extremely unlikely to recolonise a new migratory route. So we may need to bring grey whales to our shores – and this will be a challenge.

Airlifting whales is technically possible, but these are amongst the most emotionally sensitive creatures of all; grieving mothers will sometimes keep close to their dead calves for days at a time. We may not yet have the technology to move whales ethically, but we might begin to invest, now, in solving such a problem. To imagine a future where grey whales once again grace our coasts, feed our seabirds and swim placidly beside us from the Hebrides to the Gower, is something almost painfully beautiful to contemplate. Actively restoring whales will be at the hard end of what is achievable, but in recent years, rhinos have been lifted by helicopter for hundreds of miles – and successfully reintroduced into different countries. Stranger things, perhaps, have happened.

Protecting our fisheries will not only enhance our own chances of longer-term survival as a species, but it will also accelerate the return of the giants. If the land is fairly resilient, the sea is extraordinarily so. In 2008, when the Arran Community Trust, in Scotland, sealed off just two square kilometres of Lamlash Bay, developing a 'no-take zone', scallop numbers not only increased six times but they also grew dramatically in size. Just like the giant hake of the seventeenth century, marine creatures not only proliferate, but also grow larger, when human predation pressure upon them decreases to a sustainable level, and often, for just a short period of time. Elsewhere, oysters, the intestines of the sea, cleaning our waters and cycling nutrients, are being restored to places like the Humber. The oceans, of course, do not suffer the limits of the land. Whilst wolves will never arrive in Britain without active human consent, many species, from bluefin tuna to dolphins, can colonise our shores from distant waters, provided we afford them safe haven. And in some ways, this makes ocean conservation as much about creating areas where we simply do nothing, as always stepping in proactively.

Birds of prey, returning already to British skies, can rapidly recolonise if their numbers are not suppressed by illegal hunting, and provided sufficiently robust populations exist in adjacent areas. Already, there should be golden eagles gracing the Pennines, Bowland and the Yorkshire Dales, as well as far greater numbers of hen harriers. The financial and inevitable decline of intensive grouse moor management will accelerate their return, both naturally, in time, and, ideally, with reintroduction schemes to help. We simply cannot afford to live in a world where golden eagles – a true keystone species – do not hunt our hills and larger woodlands. White-tailed eagles, for the first time in living memory, now grace the Isle of Wight, and satellite-tagged birds have been shown to be travelling widely. Adaptable and long-lived, they should recolonise large wetlands, such as the Somerset Levels, within a generation, but the spread of birds of prey has always been aided, never hindered, by further reintroductions, especially, for this species, in areas like the East Anglian coastline. Many other species are naturally increasing, especially goshawks, and

others, such as peregrines, have even moved into and
thrived within our cities. For many birds of prey, however,
we forget the impacts further down the trophic cascade.
Rodenticides damage vital populations of mice and voles
eaten by kestrels; in time, we must accept that these smaller
aerial predators are the best means of controlling both. But
with most of the public charmed and invested in the return
of birds of prey, it is likely that eagles will once again begin
to dominate our skies in the decades to come. The real
question is, whether we, as a species, can thrive too.

We have the knowledge, the capacity, and at times the
wisdom, to become the ultimate steward of stewards: to
mend our own actions as the ecosystem engineer in chief,
whilst ceding power, deliberately, to others. In some parts
of the world, such as Canada, this balance has already been
well struck. On our denuded island, tightly ordered and
controlled, there is no denying that we have a very long way
to go. Yet if we look at our own track record in the past 200
years, much has been achieved that was far more improbable
than rebuilding the natural world around us – and returning
our island's rightful fellow stewards. We have all the ingenuity
within us, the technology and, increasingly, the appreciation
of the necessity, to make this happen.

There will be little room for error in the ever-harder
years to come. As our climate grows ever more unpredictable,
and with it our food security, as sea levels rise and extreme
weather becomes increasingly the norm, we must look to
our fellow species, as much as to ourselves, to solve the
problems we've created. And in having the humility to share,
to cede power and to allow other wild forces to work with
us and beside us, we will, once again, become – for us and
for others – a cornerstone species to stand the test of time: a
rightful steward of the land.

Acknowledgements

I would like to thank the many inspirational individuals and organisations who have made this book possible through their meticulous study of keystone species, their habitats, threats, ecological history and ecosystem effect. These include:

S. Elizabeth Alter, Henrik Andrén, Marie Baltzinger, Ross Barnett, Lucas Battisti, Alison Benjamin, Robert L. Beschta, Andrzej Bobiec, Richard Brazier, Gabe Brown, John Burnside, Charles Burrell, Róisín Campbell-Palmer, Sean Carroll, Rachel Carson, Hilary A. Cooke, Jim Crumley, Lutz Dalbeck, Charles Darwin, María del Mar Delgado, Roy Dennis, Patrick Duncan, James A. Estes, Richard Evans, Jean-Martin Fortier, Martin Gaywood, Ben Goldfarb, Ben Goldsmith, Dave Goulson, Derek Gow, Stephen J. G. Hall, Vladimír Hanzal, David Hetherington, Fernando Hiraldo, Pawel Janiszewski, Jaime E. Jiménez, Carol A. Johnston, Chris Jones, Robert Kenward, Trish J. Lavery, Alan Law, Olof Liberg, Leo Linnartz, Rui Lourenço, Brian McCallum, Renée Meissner, Wojciech Misiukiewicz, George Monbiot, Marianne Pasanen-Mortensen, Anna-Katharina Mueller, Ian Newton, John Odden, Vincenzo Penteriani, Andrew J. Pershing, Krzysztof Pietrasz, Benno Pokorny, Michael M. Pollock, Alan Puttock, James Rebanks, William J. Ripple, Stephen Roberts, Robert C. Rocha, Ana S. L. Rodrigues, Joe Roman, Irina I. Rotenko, Francisco Sánchez-Bayo, Frédéric de Schaetzen, Fabrizio Sergio, Vadim Sidorovich, Natasha K. E. Sims, Merlin Sheldrake, David J. Slater, Victor Smetacek, Laura V. Smith, Douglas W. Smith, Cameron E. Stevens, Erik Stokstad, Andrew Stringer, Isabella Tree, Deborah Uchida, Franz Vera, Jeff Watson, Nigel J. Willby, Sophie-lee Williams, Peter Wohlleben, Ben A. Woodcock, Kris A. G. Wyckhuys, Derek Yalden and Dorothy Yamamoto.

Further Reading

Chapter 1: Boar

Baltzinger, M., Mårell, A., Archaux, F., Pérot, T., Leterme, F. and Deconchat, M. 2016. Overabundant ungulates in French Sologne? Increasing red deer and wild boar pressure may not threaten woodland birds in mature forest stands. *Basic and Applied Ecology*, 17(6).

de Schaetzen, F., van Langevelde, F. and WallisDeVries, M. F. 2017. The influence of wild boar (*Sus scrofa*) on microhabitat quality for the endangered butterfly *Pyrgus malvae* in the Netherlands. *Journal of Insect Conservation*, 22: 51–59.

Macdonald, B. 2019. *Rebirding: Restoring Britain's Wildlife*. Pelagic Publishing, Exeter.

Pokorny, B. and Jelenko, I. 2013. Ecological importance and impacts of wild boar (*Sus scrofa* L.). *Zlatorgov Zbornik*, 2(2): 2–30.

Sims, N. K. E. 2005. The ecological impacts of wild boar rooting in East Sussex. http://www.britishwildboar.org.uk/The%20ecological %20impacts%20of%20wild%20boar%20rooting%20in%20East %20Sussex.pdf *(accessed 8 November 2021)*.

Tree, I. 2018., *Wilding: The Return of Nature to a British Farm*. Picador, London.

Vera, F. 2000. *Grazing Ecology and Forest History*. CABI Publishing, Wallingford.

Yamamoto, D. 2017. *Wild Boar*. Reaktion Books, London.

Chapter 2: Birds of Prey

Dennis, R. 2021. *Restoring the Wild: Sixty Years of Rewilding Our Skies, Woods and Waterways*. William Collins, Glasgow.

Evans, R., O'Toole, L. and Whitfield, D. 2012. The history of eagles in Britain and Ireland: An ecological review of placename and documentary evidence from the last 1500 years. *Bird Study*, 59(3): 1–15.

Kenward, R. 2006. *The Goshawk*. T & AD Poyser, London.

Lourenco, R., Santos, S. M., Rabaça, J. E. and Penteriani, V. 2011. Superpredation patterns in four large European raptors. *Population Ecology*, 53: 175–185.

Mueller, A., Chakarov, N., Heseker, H. and Krüger, O. 2016. Intraguild predation leads to cascading effects on habitat choice, behaviour and reproductive performance. *Journal of Animal Ecology*, 85(3): 774–84.

Newton, I. 1987. *The Sparrowhawk*, Shire Publications. London.

Newton, I. 1998. *Population Limitation in Birds.* Academic Press, Cambridge, Massachusetts.

Newton, I. 2010. *Population Ecology of Raptors.* T & AD Poyser, London.

Penteriani, V. and Delgado, M. 2019. *The Eagle Owl.* T & AD Poyser, London.

Sergio, F. and Hiraldo, F. 2008. Intraguild predation in raptor assemblages: a review. *IBIS*, 150(1): 132–145.

Watson, J. 2010. *The Golden Eagle.* T & AD Poyser, London.

Williams, S., Perkins, S. E., Dennis, R., Byrne, J. P. and Thomas, R. J. 2020. An evidence-based assessment of the past distribution of Golden and White-tailed Eagles across Wales. *Conservation Science and Practice*, 2(2).

Yalden, D. 2007. The older history of the White-tailed Eagle in Britain, *British Birds*, 100(8): 471–480.

Chapter 3: Beavers

Brazier, R., Puttock, A., Graham, H. A., Auster, R. E., Davies K. H. and Brown, C. M. L. 2020. Beaver: Nature's ecosystem engineers. *WIREs Water*, 8(1).

Campbell-Palmer, R., *et al.* 2015. *The Eurasian Beaver*, Pelagic Publishing, Exeter.

Campbell-Palmer, R., *et al.* 2016. *The Eurasian Beaver Handbook: Ecology and Management of* Castor fiber, Pelagic Publishing, Exeter.

Cooke, H. A. 2008. Influence of beaver dam density on riparian areas and riparian birds in shrubsteppe of Wyoming. *Western North American Naturalist*, 68: 365–373.

Crumley, J. 2015. *Nature's Architect: The Beaver's Return to Our Wild Landscapes*, Saraband, Manchester.

Dalbeck, L., Hachtel, M. and Campbell-Palmer, R. 2020. A review of the influence of beaver *Castor fiber* on amphibian assemblages in the floodplains of European temperate streams and rivers. *The Herpetological Journal*, 30(3): 135–146.

Dalbeck, L., Lüscher, B. and Ohlhoff, D. 2007. Beaver ponds as habitat of amphibian communities in a central European highland. *Amphibia-Reptilia*, 28(4): 493–501.

Goldfarb, B. 2019. *Eager: The Surprising, Secret Life of Beavers and Why They Matter.* Chelsea Green Publishing, Chelsea, Vermont.

Gow, D. 2020. *Bringing Back the Beaver: The Story of One Man's Quest to Rewild Britain's Waterways.* Chelsea Green Publishing, Chelsea, Vermont.

Janiszewski, P., Hanzal, V. and Misiukiewicz, W. 2014. The Eurasian Beaver (*Castor fiber*) as a Keystone Species – a Literature Review, *Baltic Forestry*, 20(2): 277–286.

Author note: This article constitutes, in the author's opinion, one of
the most comprehensive single syntheses of the beaver's ecosystem
impacts on individual species and faunal and floral groups in Europe,
drawing on a wide range of studies across Europe. Studies mentioned
in the text, and not specifically cited, can be found cited in the
paper above.

Jiménez, J. E. 2012. Do beavers improve the habitat quality for Magellanic
Woodpeckers? *Bosque*, 33(3): 271–274.
Johnston, C. A. 2017. *Beavers: Boreal Ecosystem Engineers*, Springer, London.
Kemp, P., Worthington, T. A., Langford, T. E. L. and Tree, A. 2011.
Qualitative and quantitative effects of reintroduced beavers on stream
fish. *Fish and Fisheries,* 13(2): 158–181.
Law, A., Gaywood, M. J., Jones, K. C., Ramsay, P. and Willby, N. J. 2017.
Using ecosystem engineers as tools in habitat restoration and
rewilding: beaver and wetlands. *Science of The Total Environment*, volumes
605-606: 1021–1030.
Law, A., Levanoni, O., Foster, G., Ecke, F. and Willby, N. J. 2019. Are beavers
a solution to the freshwater biodiversity crisis?. *Diversity and
Distributions*, 25(11): 1763-1772.
Pietrasz, K., Sikora, D., Chodkiewicz, T., Ślęzak, M. and Woźniak, B.
2019. Keystone role of Eurasian beaver *Castor fiber* in creating the
suitable habitat over the core breeding range for forest specialist
species the three-toed woodpecker *Picoides tridactylus*. *Baltic Forestry*,
25(2).
Pollock, M. M., Pess, G. R. and Beechie, T. 2004. The Importance of Beaver
Ponds to Coho Salmon Production in the Stillaguamish River Basin,
Washington, USA. *North American Journal of Fisheries Management*, 24(3):
749–760.
Stevens, C. E., Pazkowski, C. A. and Foote, L. 2007. Beaver (*Castor
canadensis*) as a surrogate species for conserving anuran amphibians on
boreal streams in Alberta, Canada. *Biological Conservation*, 134(1): 1–13.
Stringer, A.P., Blake, D. and Gaywood, M. J. 2015. A review of beaver
(*Castor* spp.) impacts on biodiversity, and potential impacts following a
reintroduction to Scotland. *Scottish Natural Heritage Commissioned
Report*, No. 815.

Chapter 4: Whales

Alter, S. E., Rynes, E. and Palumbi, S. R. 2007. DNA evidence for historic
population size and past ecosystem impacts of gray whales. *PNAS*,
104(38): 15162–15167.
Estes, J. A., Doak, D. F., Springer, A. M. and Williams, T. M. 2009. Causes
and consequences of marine mammal population declines in

southwest Alaska: a food-web perspective. *Philosophical Transactions of the Royal Society B*, 364(1524), 1647–1658.

Lavery, T. J., Roudnew, B., Seymour, J., Mitchell, J. G., Smetacek, V. and Nicol, S. 2014. Whales sustain fisheries: Blue whales stimulate primary production in the Southern Ocean. *Marine Mammal Science*, 30(3): 888–904.

Pershing, A. J., Christensen, L. B., Record, N. R., Sherwood, G. D. and Stetson, P. B. 2010. The Impact of Whaling on the Ocean Carbon Cycle: Why Bigger Was Better. *PLoS ONE*, 5(8).

Rocha, R. C., Jr., Clapham, P. J. and Ivashchenko, Y. 2015. Emptying the Oceans: A Summary of Industrial Whaling Catches in the 20th Century. *Marine Fisheries Review*, 76(4): 37–48.

Rodrigues, A. S., Charpentier, A., Bernal-Casasola, D., Gardeisen, A., Nores, C., Millán, J. A. P., McGrath, K. and Speller, C. F. 2018. Forgotten Mediterranean calving grounds of grey and North Atlantic right whales: evidence from Roman archaeological records. *Proceedings of the Royal Society B*, 285(1882).

Roman J. and Palumbi, S. R. 2003. Whales Before Whaling in the North Atlantic. *Science*, 301(5632), 508–510.

Roman, J., Estes, J. A., Morisette, L., Smith, C., Costa, D., McCarthy, J., Nation, J. B., Nicol, S., Pershing, A. and Smetacek, V. 2014. Whales as marine ecosystem engineers. *Frontiers in Ecology and the Environment*, 12(7): 377–385.

Smetacek, V. 2006. *Are declining Antarctic krill stocks a result of global warming or decimation of the whales?* Invited lecture, Scientific Debate on Impacts of Global Warming on Polar Ecosystems, BBVA Foundation, Palacia de Marqués de Salamanca, Madrid.

Smith, L. V., McMinn, A., Martin, A., Nicol, S., Bowie, A. R., Lannuzel, D. and van der Merwe, P. 2013. Preliminary investigation into the stimulation of phytoplankton photophysiology and growth by whale faeces. *Journal of Experimental Marine Biology and Ecology*, 446: 1–9.

Chapter 5: Bees

Battisti, L., Potrich, M., Sampaio, A. R., de Castilhos Ghisi, N., Costa-Maia, F. M., Abati, R., dos Reis Martinez, C. B. and Sofia, S. H. 2021. Is glyphosate toxic to bees? A meta-analytical review. *Science of The Total Environment*, volume 767.

Benjamin, A. and McCallum, B. 2009. *A World without Bees,* Guardian Books, London.

Carson, R. 2000. *Silent Spring*. Penguin Classics, London.

Darwin, C. 1859. *On the Origin of Species by Means of Natural Selection*. John Murray, London.

Goulson, D. 2009. *Bumblebees: Behaviour, Ecology and Conservation*. Oxford University Press, Oxford.

Goulson, D. 2015. *A Buzz in the Meadow*. Vintage, London,

Goulson, D. 2020. *The Garden Jungle: or Gardening to Save the Planet*. Vintage, London.

Goulson, D. 2021. *Silent Earth: Averting the Insect Apocalypse*. Jonathan Cape, London.

Snchez-Bayo, F. and Wyckhuys, K. A. G. 2019. Worldwide decline of the entomofauna: A review of its drivers. *Biological Conservation*, 232: 8–27.

Woodcock, B. A., *et al.* 2016. Impacts of neonicotinoid use on long-term population changes in wild bees in England. *Nature Communications*, 7.

Woodcock, B. A., *et al.* 2017. Country-specific effects of neonicotinoid pesticides on honey bees and wild bees. *Science*, 356(6345): 1393–1395.

Chapter 6: Cattle and Horses

Burnside, J. 2021. *Aurochs and Auks: Essays on Mortality and Extinction*. Little Toller Books, Beaminster.

Duncan, P. 1991. *Horses and Grasses: The Nutritional Ecology of Equids and Their Impact in the Carmargue*, Springer, London.

Hall, S. J. G. 2008. A comparative analysis of the habitat of the extinct aurochs and other prehistoric mammals in Britain, *Ecography*, 31(2): 187–190.

Linnartz, L. and Meissner, R. 2014. *Rewilding Horses in Europe: Background and guidelines – a living document*. Rewilding Europe, Nijmegen.

Stokstad, E. 2015. Bringing back the Aurochs. *Science*, 350(6265): 1144–1147.

Tree, I. 2018. *Wilding: The Return of Nature to a British Farm*. Picador, London.

Uchida, D. 2019. *Wild Horses of Mongolia: The Przewalski Horse in Hustai National Park*. Kindle edition.

Vera, F. 2000. *Grazing Ecology and Forest History*. CABI Publishing, Wallingford.

Chapter 7: Trees

Bobiec, A., *et al.* 2005. *The Afterlife of a Tree*. WWF Poland, Białystok.

Sheldrake, M. 2021. *Entangled Life: How Fungi Make Our Worlds, Change Our Minds and Shapre Our Futures*. Vintage, London.

Wohlleben, P. 2017. *The Hidden Life of Trees: What They Fell, How They Communicate: Discoveries from a Secret World*. William Collins, Glasgow.

Wohlleben, P. 2018. *The Secret Network of Nature: The Delicate Balance of All Living Things*. Vintage, London.

Chapter 8: Lynx and Wolves

Andrén, H. and Liberg, O. 2015. Large Impact of Eurasian Lynx Predation on Roe Deer Population Dynamics. *PLoS ONE*, 10(3).

Barnett, R. 2020. *The Missing Lynx: The Past and Future of Britain's Mammals*, Bloomsbury Wildlife, London.

Beschta, R. L. and Ripples, W. J. 2010. Recovering Riparian Plant Communities with Wolves in Northern Yellowstone, U.S.A. *Restoration Ecology* 18(3): 380–389.

Carroll, S.B. 2017. *The Serengeti Rules: The Quest to Discover How Life Works and Why It Matters*. Princeton University Press, Princeton.

Dennis, R. 2016. *Wildcat & Lynx – always remember the bigger picture*. https://www.roydennis.org/2016/06/17/wildcat-lynx-always-remember-the-bigger-picture/ *(accessed 21 November 2021)*.

Hetherington, D. 2018. *The Lynx and Us*. Wild Media Foundation, Inverness-shire.

Odden, J., *et al*. 2002. Lynx Depredation on Domestic Sheep in Norway, *Journal of Wildlife Management*, 66(1): 98–105.

Pasenen-Mortensen, M., Pyykönen, M. and Elmhagen, B. 2013. Where lynx prevail, foxes will fail – limitation of a mesopredator in Eurasia, *Global Ecology and Biogeography*, 22(7): 868–877.

Ripples, W. J. and Beschta, R. L. 2011. Trophic cascades in Yellowstone: The first 15 years after wolf reintroduction. *Biological Conservation*, 145(1): 205–213.

Ripples, W. J., Beschta, R. L. and Painter, L. E. 2015. Trophic cascades from wolves to alders in Yellowstone. *Forest Ecology and Management*, 345: 254–260.

Sidorovich, V. E. 2006. Relationship between prey availability and population dynamics of the Eurasian lynx and its diet in northern Belarus. *Mammal Research*, 51: 265–274.

Sidorovich, V. E. and Rotenko, I. 2019. *Reproduction biology in grey wolves* Canis lupus *in Belarus: Common beliefs versus reality*. Chatyry Chverci, Minsk.

Author note: The core studies cited can be found on Sidorovich's blog: https://sidorovich.blog/ (accessed: 21 November 2021). In the author's opinion, these are the most revelatory field studies of wolves and lynx in Europe, in a comparable environment to that we once had here in the UK.

Sidorovich, V. E., Gouwy, J. and Rotenko, I. 2018. *Unknown Eurasian lynx* Lynx lynx: *New findings on the species ecology and behaviour,* Chatyry Chverci, Minsk.

Sidorovich, V. E., Rotenjo, I. I. and Krasko, D. A. 2011. Badger *Meles meles* Spatial Structure and Diet in an Area of Low Earthworm Biomass and High Predation Risk. *Annales Zoologici Fennici*, 48(1): 1–16.

Smith, D. W., *et al.* 2020. *Yellowstone Wolves: Science and Discovery in the World's First National Park*. University of Chicago Press, Chicago.

Smith, D. W., Peterson, R. O. and Houston, D. B. 2003. Yellowstone after Wolves. *BioScience*, 53(4): 330–340.

Chapter 9: Humans

Brown, G. 2018. *Dirt to Soil: One Family's Journey into Regenerative Agriculture*. Chelsea Green Publishing, Chelsea, Vermont.

Fortier, J-M. 2014. *The Market Gardener: A Successful Grower's Handbook for Small-Scale Organic Farming*. New Society Publishers, Gabriola Island, British Columbia.

Rebanks, J. 2021. *English Pastoral: An Inheritance*. Penguin, London.

Index